SPACE AS A STRATEGIC ASSET

Space as a Strategic Asset

Joan Johnson-Freese

COLUMBIA UNIVERSITY PRESS  New York

Columbia University Press
Publishers Since 1893
New York Chichester, West Sussex

Library of Congress Cataloging-in-Publication Data
Johnson-Freese, Joan.
Space as a strategic asset / Joan Johnson-Freese.
p. cm.
Includes bibliographical references and index.
ISBN 0-231-13654-4 (hardback : alk. paper)
ISBN 0-231-51001-2 (e-book)
1. Outer space—Strategic aspects. 2. Space weapons.
3. Outer space—Government policy—United States. I. Title.

JZ5695.J64 2007
358'.8—dc22
2006031631

Columbia University Press books are printed
on permanent and durable acid-free paper.
Printed in the United States of America
c 10 9 8 7 6 5 4 3 2 1

Contents

Preface

The United States is currently the preeminent global space power. American space technology provides applications on earth, such as communications and navigation, that improve everyday lives and advance economic development. The U.S. aerospace industry, the global leader, is a lucrative industry. Some sectors are expanding due to increased global demands for products; others, specifically communications satellites, are highly restricted compared with their global counterparts. Space technology also provides tools for American warfighters, giving U.S. troops heretofore unimaginable advantages. Finally, space technology allowed the United States perhaps its most inspiring global moment, when Neil Armstrong first walked on the moon in 1969. That "one small step for a man, one giant leap for mankind" symbolized U.S. leadership and its future-oriented direction. None of those advances or advantages, however, can be taken for granted. They must be nurtured and protected. The premise of this book is that many current space policies are neither nurturing nor protective, and must be reconsidered. Too often and too many policies are stovepiped, contradictory, or impossible to execute, not always based on careful analysis, and produce unintended consequences that can damage U.S. space preeminence, to the overall detriment of the country. Reconsideration of these policies must be broad-based, not just carried out by those who created the ill-begotten policies in the first place. This book attempts to provide the background and information to allow that broad-based reconsideration to occur.

Unfortunately, a number of important issues associated with strategic space issues are beyond full examination here. Orbital debris, for example, already affects space operations and will be an increasing issue in the future. One of the first theories about the space shuttle *Columbia*'s catastrophic fall from the sky was that it was hit by space debris. Yet, much like nuclear waste disposal, the issue has simply been set aside until an undefined later date, because it could be. When a piece of space debris the size of a quarter does catastrophic damage to a spacecraft, we will deal with it then. Another key issue to future U.S. space policy is the projected shortfall in the science and engineering workforce, and the corollary issue of U.S. science and engineering graduate programs being increasingly populated by non-American students.[1] While worthy of full investigation, that is beyond the scope of this book. Issues such as the technical options for and cost-benefit ratios of "hardening" U.S. space assets to afford them further protection are not explored either. And, unfortunately, full examinations of the space programs of many nations, such as Russia, India, Brazil, and Japan, while fully warranted and undoubtedly very useful, are not included. Every attempt is made to cite examples from a variety of countries and acknowledge as many of the multiple omitted issues as possible, but this book remains focused on those strategic issues related to the security dilemma created by the potential clash of global space ambitions, and how the United States can fulfill its destiny as a great, flourishing, and powerful nation without such a clash.[2]

The technical material inherently part of space issues is presented as nontechnically as possible. Technical terms can and have intimidated the public (sometimes deliberately) from participating in debates on issues they feel they cannot understand. Therefore, many military program names, which often include four or five adjective descriptors, and military jargon are simplified or omitted, as are technical terms. That will inherently result in some simplifications or restatements likely to irritate those who are intimately familiar with complex technical programs and issues. Such simplification is necessary, however, to bring a wider audience into the discussion. The overall intent is to provide a comprehensive overview of the issues, to allow and promote discussion and debate.

The issues raised in this book are too important to remain opaque to the public and politicians. Meaningful appreciation, however, requires an understanding of the basics of these complex issues, which in some instances actually involve rocket science. In chapters 2 to 8, I attempt to explain the complex issues involved, in many cases including the historical context and simple explanations of the technical parameters involved. Chapter 2 presents the dilemmas created by the inherently dual-use nature of space technology, because those technical dilemmas overshadow all political discussions. Chapter 3 examines the U.S. manned space program. Chapters 4 and 5 concern the military uses of space and the paths to space weaponization. Chapter 6 focuses on the state of the U.S aerospace industry and the issues created by current U.S. export control laws. Chapter 7 covers the space ambitions of developed countries, specifically as exemplified by European efforts. Chapter 8 does the same for developing countries, specifically as exemplified by China. I reserve recommendations regarding alternatives to the potentially perilous path currently being pursued by the United States until chapter 9, as it is important that the issues be considered from a unitary perspective rather than through the traditional stovepipes of military, NASA, or commercial activity.

The United States needs a comprehensive space strategy that considers all aspects of strategic activity and the realities of the international environment. Such a strategy does not yet exist. The long-awaited new National Space Policy, (finally) issued in October 2006, apparently after wrangling over some thirty-five draft versions, focuses on space from a military perspective, with manned space, export control, and other issues tangentially addressed and little regard given to the realities of either the budget or the international environment. Space, however, is a strategic asset and needs to be viewed as such. Not doing so creates an unacceptable strategic risk for the United States.

Finally, this book narrates a baseball game in progress. While every attempt is made to provide current information, events march on, new documents are issued, and people change jobs. However, until the United States comprehensively addresses the issues presented in this book, its fundamental premises remain valid and vital for consideration.

Acknowledgments

This book is the culmination of research conducted over twenty-two years of studying, researching, writing, and teaching on the subject of space policy. I will never be able to individually thank all the students, friends, colleagues, and critics—and over the years, many students have become friends, colleagues, and critics—who have informed and shaped my views, but I am certainly grateful. At the University of Central Florida, enthusiasm for space from students in my space policy classes spurred me beyond my initial forays into writing about space. At the Air War College in Alabama, I first learned about the military side of space, and learning from the inside was invaluable. Colleagues at the Institute for Space and Astronautical Science in Japan first impressed upon me that not everybody works and thinks like Americans do. Over the years and many great summers in international capitals, I have lived the "3-I" philosophy inherent in space studies—international, interdisciplinary, and intercultural—with colleagues and students at the International Space University. Added to all these must be the numerous individuals at conferences, workshops, and symposia who have commented on my work and shared their own. The space policy community is small, but tight. We love to talk to one another.

Many people have specifically provided assistance and encouragement during the writing of this book, reading drafts, commenting on chapter material drawn from previous articles, and simply discussing ideas. Chris Hoeber at Space Systems/Loral has been reading drafts for what must seem like forever to him, trying to keep me technically honest. Professor Andy Stigler at the

Naval War College provided valuable comments on strategic communication issues. Theresa Hitchens at the Center for Defense Information and Gregory Kulacki at the Union of Concerned Scientists have been great about reading and commenting on material. Professor Tom Nichols at the Naval War College has been a motivator, critic, commentator, and friend. Friends Zona Douthit and Yvette Rogers have been my "lay readers"—not letting me get away with too many acronyms or five-adjective program descriptors.

I have the privilege of writing this study as a member of the faculty at the United States Naval War College. This project benefited greatly from the ongoing support for such endeavors from the college's administration, and I wish to thank that administration for encouraging such a supportive environment.

I would also like to thank my family, who is always there to patiently support and encourage me.

Finally, I would like to dedicate this book to Dr. Hermann Strub and Dr. David Webb. Hermann Strub was my first space mentor. He introduced me to the space community, allowing me to conduct research that would otherwise not have been possible. David Webb inspired me. His enthusiasm for space as a positive road to the future is contagious. I am indebted to each, and each has my deepest gratitude.

Clearly, I have had a lot of support, encouragement, and assistance in this book project, and I am very appreciative. Just as clearly, I am fully responsible for its content, conclusions, and flaws.

SPACE AS A STRATEGIC ASSET

A Clash of Ambitions | ONE

The destiny of the United States [is] to be a great, a flourishing, and a powerful nation.
—James Madison, State of the Union Address

Between October 2003 and October 2004, four space-related events occurred with varying degrees of public and media notice, ranging from none to modest. First, on October 15, 2003, China successfully orbited its first astronaut, or *taikonaut*,[1] in the *Shenzhou V* capsule,[2] joining the United States and Russia in the exclusive club of countries capable of manned spaceflight. Later that month, on October 30, China and the European Union (EU) signed an agreement making China a stakeholder in the European Union's Galileo navigation satellite program, which is likely to rival the capabilities of the U.S. global positioning system (GPS). Navigation satellites facilitate everything from use of automatic teller machines to airplane safety to tourists' ability to navigate cities using the NeverLost system on their rental cars—even to American bombers dropping precision-guided munitions on Iraq. The first two events, one featuring a new manned spaceflight program in a developing country, the other focusing on the commercial value of space, occurred outside the United States.

Then, in January 2004, the Bush administration announced a program called the New Vision for Space Exploration, intended to take the United States to the moon, Mars, and beyond. While perhaps well intended, unfortunately the viability of the program has been questioned from the start. Alexander Rose called it "a scheme destined to stall," with time lines and budgets that "are masterworks of science fiction."[3]

Finally, in August 2004, the United States Air Force issued Air Force Doctrine Document 2-2.1, entitled *Counterspace Operations*. That document, perhaps more than any before it, declared the strategic ambitions of the Air Force in space. It states that space operations protecting U.S. assets have both defensive and offensive elements, an allusion to not only space weapons but potentially their preemptive use. Part of the intent of the low-key release was to see how much notice, and subsequently ire, it would draw. It went virtually unnoticed. Consequently, in May 2005, the Air Force requested specific administration approval of a directive in the new National Space Policy to "move the United States closer to fielding offensive and defensive space weapons."[4]

Separately, these events are interesting, particularly to space enthusiasts. But taken together, they indicate an alarming and prevalent trend. The interests and ambitions of the United States in space point in a different direction from that of the rest of the world: the United States is moving toward weaponizing space for both defensive and offensive purposes, while the rest of the world considers space assets primarily as tools requisite to advance in a globalized world and is fearful of apparent U.S. intentions to arm and control the heavens, potentially shutting out other countries. The strategic and geopolitical implications of this trend raise the importance of the individual events by orders of magnitude.

Few would argue against the need for military space assets. The militarization of space, referring to the military recognizing the value of and using space assets, occurred long ago—even before humans journeyed there—and has given the United States significant security advantages that must be protected in the future. The issue is how to protect those advantages. Space weapons, which had heretofore been explicitly deemed not in the U.S. interest to pursue, appear to be the Bush administration's option of choice for that mission and others. In discussions regarding the advantages and disadvantages of space weaponization, it is important to remember that those who believe that weaponization would actually diminish rather than improve the security of U.S. assets in space are not necessarily against the military use of space. The distinction between militarization and weaponization is sometimes blurred, intentionally or unintentionally. A *New York Post* editorial that decried those questioning potential movement toward weaponizing space called it "militarizing" space in its title.[5] Obscuring the issue is not useful; what is needed is a clear debate, based on evidence and analysis, about which short-term and long-term actions will be in the best interests of the United States.

At some point in the future, the United States may need space weapons to protect its national interests and the national security of the country. Personally, though, I have not yet been convinced of that. In 2003, Michael Krepon, president emeritus of the Henry L. Stimson Center, raised multiple issues in anticipation of a vigorous debate on the need and desirability of space weapons.[6] That debate has not occurred. What has occurred, however, is the slow but consistent movement by the United States down a road leading to near-term weaponization.

The movement toward weaponization has been accomplished through a carefully choreographed plan intended to present information to the public and lawmakers while simultaneously encouraging them to pay no attention—which has not been a difficult task. Decisions have been made, doctrine written, documents published, and technology developed in full view of Congress, the press, and the American public. Yet all seem uninterested. When the occasional media piece raising concerns does appear, few beyond those already interested read it, and weaponization advocates respond with charges of fear-mongering or "whipping up anxieties with little rational justification."[7] Perspectives clearly differ. Those against weaponization see media attention as lax, while advocates regard it as "near-hysterical ranting."[8] The general polarization of views coincides with the post–September 11 political environment, charged with fear and dividing the world into simplistic, either-or choices: us and them, warriors and wimps, patriots and Democrats, a path of fear or a path of confidence. A more nuanced approach, though admittedly more difficult, will ultimately better serve the United States.

Regardless of media attention, however, analysis and debate within the government before making decisions about space policy appear painfully thin. Ideology and an assumption that weaponization is "inevitable" trump rigorous analytic considerations. RAND analyst Karl Mueller has stated that the inevitability premise is "based on a smattering of evidence and logic, extrapolated into facile overgeneralizations that are well suited for television talk-show punditry but which are a poor basis for national decision-making."[9] Facts should matter; unfortunately, they are not currently in vogue.

Ed Crane, president of the Cato Institute, wrote that President George W. Bush did not succeed in his campaign to convince the United States to privatize part of Social Security because he had focused too much on the "green-eyeshade issues such as solvency, transition costs, unfunded liabilities and rates of return."[10] He advised the White House to make Social Security privatization "an emotional issue," steering the discussion away from the harsh glare of numbers and facts. The same tactic has prevailed in space policy as well.

Just how much is the U.S. public willing to reject fact and science? Apparently, quite a lot.[11] But while emotion and passion are a part of the American spirit, so too is a hard-nosed, sensible pragmatism that turns dreams into reality. Perhaps the weaponization advocates can present a compelling case. They should have been required to do so before the country embarked on space weaponization, and they must be compelled to do it now, before the United States proceeds any further. We have a choice in the matter, and we need to choose the option that is in the best long-term interests of the United States.

It is almost as though Congress, the media, and the public simply do not want to know about the new direction the United States is taking in space policy, perhaps because they already have too much to deal with regarding the global war on terror—designated "the long war" by Defense Secretary Donald Rumsfeld in February 2006—Iraq, Social Security, health care, and a list of other consuming concerns. Alternatively, perhaps consumerism is taking its toll, with people just too fat and happy to care. Space weapons may seem too distant, too technical, or too unimaginable to deal with. Perhaps movies have made the public think that space weapons are normal. Whatever the case, those who want to pursue weaponization have been more than happy to encourage the public to remain uninvolved, allowing weaponization advocates to avoid scrutiny and, in military jargon, to "fly below the radar." A May 2005 *New York Times* article[12] on space weaponization created a short burst of media interest. Shortly thereafter, I was asked to debate the topic on two National Public Radio programs. According to the producers for both programs, neither could get a military representative to speak on their shows to state or support the Air Force position. Saying nothing proved wise, as public interest died quickly. However, the mere potential of engendering public notice may have sufficiently motivated the White House to tone down the content of its new U.S. space policy, away from overt support for weaponization and toward retaining the ambiguity has prevailed.

But discussion and debate must occur. Space weapons, I contend, are not in the best interests of the United States, at least in the near term. Asking and then doing what is in the best interests of the United States should be, I believe, the ultimate goal of both U.S. policies and the funded programs that support those policies. Alternative paths to protecting U.S. space assets are not being considered. Regardless of what path is taken, open and public discussion and debate are key. As Walter Russell Mead argues, "If after all there is one lesson that American history seems unambiguously to teach, it is that a wide and free debate is the best means to assure the prosperity, the destiny, the liberty, and the safety of the American people."[13] If arguments in support of a position are sound, debate should be welcome. Usually, only those supporting the most ideological and partisan positions avoid debate, refusing to even consider paths of action other than their own. The United States deserves better than that. Further, the debate should not focus solely on the military aspects of space, but include other aspects as well, including the future of manned spaceflight.

Because the United States is the sole remaining superpower, the role of global strategic leadership has been thrust on it, and, thankfully, it is eminently qualified to respond. The United States not only has the military might to lead, but has been viewed, as former president Ronald Reagan put it, as a "shining city on the hill"[14] that others want to emulate. The United States went to the moon because it could. It dares to do difficult things that other nations do not dare to do. Now, however, the United States appears to be ready to cede aspects of the leadership that Reagan envisioned, including leadership in manned spaceflight. Clearly, doing so will cost the United States in its global leadership role.

It is time for the United States to comprehensively look at its future in space, without narrowing the examination to either its manned space program or its military space program. One directly and significantly affects the importance of the other in establishing and keeping U.S. global leadership. Space is too important as a strategic asset to leave its fate to inertia, apathy, or a few individuals. Let the debate begin.

Space as a Strategic Asset

In international relations, when two or more states are drawn into conflict even though none of the states actually desires conflict, it is called a security dilemma. Robert Jervis wrote of situations in which the fear of being exploited drives countries to act in ways that ultimately might not be in their best interests.[15] Unfortunately, the conditions that Jervis and others depicted concerning where and how that might occur describe the current situation regarding space to an alarming degree.[16] Consequently, the actions of international space players, particularly but not exclusively the United States and China, can be better understood through the lens of a security dilemma analysis.

In his work, Jervis models international interactions through analogies to Jean-Jacques Rousseau's stag hunt and the classic prisoner's dilemma. In the stag hunt analogy, all parties benefit if they work toward a common goal, so the idea is to motivate them to work together on that goal rather than allowing them to defect to less desirable goals that result in some parties not benefiting at all. It is doubtful, however, whether there is a common goal regarding space activity in the international community. Since September 11, the Pentagon has concluded that in an uncertain security environment, as is currently pervasive, the security of the United States depends on capabilities and adaptability. Space is both an enabler and a primary provider of critical war-fighting capabilities. In Pentagon parlance, space is a strategic asset. But that simple statement is fraught with ambiguity. Space is a strategic asset to other countries as well, though the phrase assumes an entirely different meaning. At the strategic level, space assets are used for arms control verification and as early-warning systems. Space offers capabilities for linking vast distances and gathering information, improving education, expanding medical resources, creating jobs, and monitoring and managing environmental issues, among others. These capabilities clearly are of strategic value to all countries. In the Pentagon, however, the term "strategic" is equated more and more often with "military," and space assets are becoming more and more often employed at the tactical level.

Because no single easy solution appears to be in the best interests of all countries, the prisoner's dilemma may be the better analogy. In that model, there are offensive as well as defensive incentives not to cooperate with others. The United States currently holds the strongest position in terms of military space capabilities. Maintaining that position is seen as critical, reflected in current government policy that views arms control cooperation as undesirable because it potentially restricts U.S. flexibility to build or use its military space capabilities. Further, because of the particular challenges presented by the dual-use nature of space technology—specifically, because technologies such as GPS and satellite imagery are of value to both the military and civilian communities—there has been spillover to shunning cooperation in any form with countries that the United States views as potentially menacing. If a state is predisposed to see another state as an adversary or a potential adversary, that influences determinations of intent as well. Hence, China's space program is of particular interest and concern to the United States.[17]

A major variable affecting how strongly the security dilemma operates is whether weapons and policies that protect the state—or, in this case, state assets—are exclusively defensive. In the case of space assets, the situation is doubly complicated, because whether or not an asset is a weapon can be difficult to distinguish. A high proportion of space assets are considered dual-use in that they have value to both the military and civilian sectors. Imagery satellites are neutral in themselves: whether the imagery produced is used to target weapons or to monitor crop yields determines whether it is a military or a civilian satellite. The technological ambiguity has far-reaching effects regarding differences in assumptions made about dual-use technology and when and how it should be controlled. Much of the world considers communication satellites as tools for disseminating information, education, and democracy, but U.S. policy currently considers them to be "sensitive technology," meaning technology with military value. Consequently, for export sale, they are subject to the same rules that govern weapons sales. Once all the military–civilian dual-use technology issues are sorted out, if they ever can be, the issue of whether certain military technologies are more appropriate for offense or defense must also be considered.

A missile that can target another missile in flight (missile defense) can also target orbiting satellites. In fact, the latter is technologically easier, because a satellite travels in a known orbit or is stationary, and as a bright object against a dark background, it is easier to find than a moving target. From a technical perspective, it is not difficult to conceive of a missile defense system as an offensive antisatellite (ASAT) weapon. Jervis specifically talked about antiballistic missile technology:

> A beneficial consequence of a difference between offensive and defensive weapons is that if all states support the status quo [rather than an arms race] an obvious arms control agreement is a ban on weapons that are useful for attacking. . . . The fact that such treaties have been rare—the Washington naval agreements and the anti-ABM treaty can be cited as examples—shows either that states are not always willing to guarantee the security of others, or that it is hard to distinguish offensive from defensive weapons.[18]

The Anti-Ballistic Missile Treaty exemplifies the banning of a technology useful for attack, demonstrating the ambiguous nature of the technology.

Even when technology is not the issue, other matters of intent or motivation must also be considered. Space has never been solely, or even primarily, about exploration. It has always been linked to other goals, usually related to foreign policy. During the Cold War, presidents starting with John F. Kennedy leveraged manned spaceflight to increase U.S. political prestige through technological credibility and capability. Space exploration was secondary. Other avenues for achieving prestige were considered, such as large-scale ocean desalinization programs, but they were rejected. Space symbolically represented the final frontier, offering the ultimate in technological prestige.

Today, one might ask why space should remain a priority. Why not have the United States positively assert its global leadership by spending an amount equal to that needed for a manned moon or Mars mission on biotechnology to combat infectious diseases or end hunger? There are three reasons. First, currently, such a space expedition is feasible only through government funding and sponsorship. Someday, it may be possible to fly American, Virgin, JAL, or an international conglomerate airline to the moon, but that is not the case today, because the expense is too high and the regulations too complex and restrictive (at least if the United States is involved). There are, however, privately funded efforts to attack disease. The Bill and Melinda Gates Foundation targets tuberculosis, and Rotary International focuses on polio. Numerous privately funded organizations are also fighting hunger. In fact, part of the problem is coordinating their efforts. Second, and related, other countries are already involved in space activity, and space activity inherently carries security implications with it. It behooves the most powerful space player, the United States, to try to shape those efforts in a peaceful manner. Finally, space travel has strong futurist connotations, and it is in the best interests of the United States to be seen as the leader into a positive future, as it was during the Apollo program.

The Apollo program, initiated by National Aeronautics and Space Administration (NASA) in 1960 and extending until 1972,[19] was a highly visible space effort to peacefully respond to Russian space achievements, but U.S. military space programs of equal importance were also under way. In the 1960s, the military was busy improving missile launch technology while developing spy satellites to better monitor a closed Soviet regime. Both the government and the private sector worked to develop communication satellite technology to connect the world. These three general types of space efforts—military, information technology (IT), and exploration—still prevail. In fact, the security dilemma is created because the first two have become so important, as countries consider space technology essential to extending and maintaining national power. For the United States, space power is defined in military terms. But power has economic and political aspects as well. Military power is merely the most obvious and, of late, the most heavily relied on by the United States.

The militaries in Germany, the United States, and the former Soviet Union first built rocket hardware during and after World War II. The rocket hardware was later adapted for space launches. In this context, the U.S. military's current interest in space, and recognition of its battlefield benefits, is neither new nor surprising. The escalation of the security dilemma created by U.S. determination to expand its domination of space, however, deserves close scrutiny. As Jervis pointed out, countries can seriously underestimate the degree to which their capabilities menace others.[20] There are also incentives to further expand those capabilities, and U.S. policy makers must consider whether that expansion is ultimately likely to be effective or counterproductive in making the United States more secure. Expanding U.S. military capabilities has been assumed to be beneficial. However, the issue has not been analytically considered, and certainly not within a context of concurrent, underfunded, low-priority manned spaceflight efforts.

The history of space activity for and by humans is relatively short but, by its very nature, complicated. It is marked by mixed motivations, developmental anomalies, interdisciplinary requirements, organizational compartmentalization, and international cooperation and competition issues. Those issues have only been exacerbated over the years as space technology has matured and its wide-ranging value recognized.

First, adding the capabilities of space assets to conventional military hardware can significantly increase the overall capability of the hardware, making space assets what is known as force multipliers; in military parlance, improving hardware renders "force enhancement" capabilities. Adding global navigation technology to missiles can dramatically increase precision. That usage turns space hardware into a tactical military asset. This technical reality has amplified the military voice in broader policy discussions generally, and space policy discussions specifically, at least in the United States.

Second, this mounting U.S. focus on the value of space for military power has occurred as the rest of the world has recognized the key role that space technology plays in national economic development. Reasoning in Europe since the 1960s has been that space programs lead to technology development, technology to industrialization, and industrialization to economic development. What industrialization was to development in the 1960s, knowledge is in the new millennium. A 2003 EU report states that space technology is vital to develop and maintain a knowledge-based society.[21] Much of the world sees the benefits that can be reaped from space assets regarding information technology, and the connectivity provided by that technology, as an integral aspect of full citizenship in a globalized world, and, therefore, access to those benefits is a national security issue. Clearly, this view differs from that of the United States regarding the national security role of space, which is based primarily on the benefits that space can provide to the military.

Further, the Bush administration seems to accept the view—promulgated in the report of the 2001 space commission chaired by Rumsfeld—that space will "inevitably" join air, land, and sea as the "fourth battlespace." Accepting that assumption requires the United States to prepare for it. Space assets are then viewed as essential to U.S. national security, not only as force multipliers, but to provide "space control," defined as "combat and combat support operations to ensure freedom of action in space for the United States and its allies and, when directed, deny an adversary freedom of action in space."[22] Whether space control also extends to force application—a more blatant path to space weapons—was initially left unspoken, was then treated ambiguously, and is increasingly becoming a de facto reality, possible because there is no technical difference between offensive and defensive weapons. Certainly Congress likes to think that the United States maintains a "no space weapons" policy, but that may be wishful thinking on its part.

With U.S. national security pegged to space, protecting space assets becomes imperative, and stopping the spread of any technology that could potentially threaten U.S. space assets is part of that imperative. From that perspective, the larger the percentage of total space assets controlled by the United States, the better. The security dilemma arises because other countries are simultaneously seeking to increase their stakes in space, for the economic benefits to be derived and, not surprising, as a part of standard military modernization efforts, at least partly in reaction to U.S. dominance. The ambitions of other countries are fueled by the desire to improve communications capabilities, explore opportunities for economic development, address a perceived technology gap, and mitigate concerns about U.S. ability to execute unconstrained policies in space.

In general, however, space has become a military priority for the United States, while primarily a developmental and commercial priority for everyone else. In a security dilemma in which weapons are vulnerable and generate use-or-lose attitudes, and in which there is no distinction between offensive and defensive weapons, a zero-sum approach prevails. Commercial transactions are not necessarily zero-sum; transactions can satisfy both buyers and sellers. But as space is zero-sum for the United States, if other countries subsequently perceive their sovereignty to be threatened, it will soon become zero-sum for them as well. Perhaps it already has for China, with its ambitious space program and growing inventory of inherently dual-use technology.

The Launch of *Shenzhou V*

When China launched then-Lieutenant Colonel Yang Liwei into space on October 15, 2003, it became only the third country capable of manned spaceflight. Celebrations ensued in China, albeit mostly choreographed by the government. Asian governments and publics paid considerable attention; European governments were more interested and laudatory than was the European public. Meanwhile, in the United States, Yang had a hard time competing with the baseball playoffs for public attention, though the government offered restrained congratulations. While the internal applause and external attention generated by Yang's flight were appreciated, they hardly seem enough to motivate a program costing over $2 billion to a country of more than 1.3 billion people, with Herculean issues providing water, food, housing, and employment to its population.

China has several motivations for undertaking Project 921, its manned space program, including economic development, creating jobs, and building dual-use technology useful for the military. Clearly, however, China also seeks the same kind of external prestige and domestic credibility that the United States reaped from the Apollo program. The prestige garnered by the launch, and the successful demonstration of its technical capabilities in manned spaceflight, carry significant geopolitical implications. Especially in today's globalized environment, technology advancements can be viewed to indicate national stature and, potentially, power. Techno-nationalism—using technology to build stature and power perceptions—is a useful and valid geopolitical consideration, as the U.S. Apollo program demonstrated.

Regionally, China's role is changing. Since 2000, China has launched a major regional charm campaign involving aid to needy countries and establishing beneficial trade relationships. The campaign has been largely successful (with the key exceptions of Taiwan and Japan) in transforming China's image from regional bully to regional power. The timing of the campaign also coincides with taking advantage of what China sees as a U.S. obsession with terrorism. Its success is evidenced in poll data. A Pew Research Center poll taken in April and May 2005 showed that China, a Communist dictatorship, was viewed more favorably than was the United States in eleven of the sixteen countries surveyed: Britain, France, Germany, Spain, the Netherlands, Russia, Turkey, Pakistan, Lebanon, Jordan, and Indonesia. India and Poland saw the United States in a more favorable light than they did China, and Canada was about evenly split.[23]

China's changed image is important. Bullies are kept at bay, while powers can be partners, even desirable ones. A space launch will not dispel political rivalries and historical feelings of distrust, but a demonstration of advanced technology can have strategic and market impact. Also, an Asian country joining two Western countries in a previously exclusive technical accomplishment can become a "relative" source of pride, especially considering the size of the Chinese diaspora throughout Asia and the world. Unquestionably, China gained regional status from the *Shenzhou V* launch.

Internationally, the Chinese launch triggered considerable geopolitical posturing regarding potential new partnerships, at many levels and on many topics. On the day after the launch, President Vladamir Putin of Russia sent a telegram to President Hu Jintao of China:

> Please accept our most sincere congratulations in connection with the historic event in China's life—the first space flight of a Chinese cosmonaut. This is a worthy and weighty outcome of the efforts that the people of China have been making for many years, and of your country's successful advancement along the road of comprehensive development and transformation of your country into a modern state of worldwide dimension. We are confident that China's full-fledged membership of the family of space powers will serve the cause of securing peace, security and stability on Earth, development of science and technology, and progress of planet Earth's civilization. Russian-Chinese space cooperation is an important trend in bilateral relations. It is making progress, it has good prospects for the future and, undoubtedly, it will bear more fruit for the benefit of our nations. Please pass our congratulations and good wishes to all those who contributed to the project to build a manned spacecraft and, of course, to the first Chinese cosmonaut.[24]

In referencing China as a modern state and a full-fledged member of the international space community, Putin was signaling a clear invitation for extended space cooperation.

On launch day, the director-general of the European Space Agency (ESA) sent warmest congratulations to China: "This mission could open a new era of wider cooperation in the world's space community."[25] The implications of that statement are significant, indicating lateral strategic issues. Europe's willingness in 2005 to consider lifting its arms embargo against China indicates its long-term view of Sino-European relations.

The United States, however, was more reserved about congratulating a Communist government and welcoming China into the international family of space powers. The United States did not want to imply a willingness to include China in other cooperative space activities, such as the International Space Station (ISS), and it has considerable trepidation about Beijing's intended ends for the dual-use technology developed in conjunction with Project 921. However, because other countries view the strategic importance of space differently from the United States, American reticence to offer China invitations for future cooperation in space was clearly not shared by others.

Previously, countries interested in manned space programs had little choice but to work with the United States or Russia. In the 1990s, with the merger of the U.S. manned program and the near-bankrupt Russian manned program in support of the ISS, the United States continued its de facto lead in manned spaceflight, which Apollo had established earlier. With China's entry into manned spaceflight, working with the United States is no longer the only option, or even the best deal. Though characterized largely as successful by the participants, particularly the United States, the ISS experience has often proved painful. The United States does not accept partners in the sense of shared or often even consultative decision making. It allows only participation, and participants sometimes grumble about being treated merely as subcontractors, making further cooperation less attractive. Additionally, U.S. export-control laws since the late 1990s have been obtuse and restrictive. Countries have increasingly avoided working with the United States on aerospace programs for fear of having them unexpectedly blocked by ambiguous U.S. regulations, after having committed scarce funds. More broadly, closer ties with China benefit countries such as France and Germany, not only for the potential lucrative market and space program options it offers, but as a potential counterweight to American power, seen as increasingly unilateral since Operation Iraqi Freedom (OIF).[26]

Galileo

Galileo, a European radio-navigation satellite program, is an example of international cooperation to counterbalance the American aerospace advantage. The GPS network and the Internet are the first global utilities, in that they provide increasingly essential services for everyone. As globalization looks to be the dominant trend of the twenty-first century, its forces are ignored at great peril. It is simultaneously an integrative force, creating networks that draw individuals, organizations, and nations closer together, and a fragmenting force that drives them apart. The current iteration of globalization, and this is at least the third,[27] is differentiated primarily by the role of information technology, which increases both the influence of networks and the speed of change. Navigation satellites are a key element of information technology through their multiple and various uses.

While Europeans have enjoyed free access to the U.S. GPS network, they have been anxious about relying on the largesse of the U.S. military, as the owner-operator of the GPS, for continued, uninterrupted service. In the mid- to late 1990s, the dominant European concern was that free access to the GPS was part of a U.S. "bait-and-switch" plan; in other words, the United States intended to hook users, and then start charging for the services. More recently, concerns are not over the cost of services, but potential denial of services. Although the Pentagon rejected a December 2004 media report that President Bush had ordered plans to temporarily disable the U.S. GPS system during a national crisis to prevent terrorists from using the system,[28] global (and domestic) fears about military ownership of such an important global utility were fueled. In fact, that same month, President Bush updated the nation's GPS policy, giving the Department of Transportation authority equal to that of the Pentagon on the committee that manages the GPS technology and spectrum. The change was at least partly to allay civilian concerns at home and abroad about GPS access. But for the many countries that consider keeping up with globalization to be a strategic goal, depending on the military-owned system of a foreign country for strategic applications—particularly a country increasingly viewed as having proclivities toward unilateralism—is not without significance. Further, dependence decreases a country's or a region's chances of maintaining any semblance of a voice in world affairs.

Beyond the value of the utilities and the desire to avoid dependence on the United States in a strategic area, the market for navigation satellite services provides compelling motivation for involvement. As a 1995 RAND report presented to the European Commission in 1999 as rationale for supporting Galileo concluded,

> The stakes are both economic and industrial. In particular, with a potential global market valued at EUR 40 thousand million between now and 2005, the challenge is to capture a fair share of the satellite navigation market as well as the jobs which flow from it. The current estimates are as follows: the development of the Galileo infrastructure would generate 20,000 jobs, while its operation would create 2,000 permanent jobs, not including opportunities in the field of applications.[29]

Through the years, estimates have only grown on the expected returns from the equipment and services market, and on the number of jobs that can be expected from the program.[30] Potential economic return, coupled with concerns about dependence on a U.S. system, proved powerful incentives. As a consequence, Europe gathered the political will to do what it had been unable to do before, and what the United States had been counting on its being unable to do: pull together the previously competing turf interests of the EU and the ESA to support the Galileo program.

In June 2004, the United States and Europe signed an agreement ending a dispute more than five years old over the potential of the European network's signals to interfere with GPS signals intended for U.S. military use. According to the agreement, Europe will shift problematic frequencies to ones that allow either side to effectively jam the other's signal in a small area, such as a battlefield, without shutting down the entire system. Equally important from the civilian perspective is that the agreement allows the systems to work interchangeably, greatly benefiting manufacturers, service providers, and consumers. Without the agreement, separate hardware and software might have been required to access Galileo and the GPS. If Galileo ends up being more commercial user–friendly than GPS, as it is intended to be, then without an agreement on compatibility, the GPS could become obsolete for civilian use, the equivalent of what VHS technology did to Betamax in the 1980s, with significant economic losses for many U.S. companies. The deployment of Galileo in itself will end the virtual U.S. commercial and military monopoly in navigation satellites.[31]

Not only was Europe able to pull together to end the U.S. navigation satellite monopoly, but it invited other countries to join the efforts as well. In October 2003, China and the European Union signed an agreement officially making China a partner in developing the Galileo program. China committed approximately $259 million in hard currency to work with EU countries on both technical and manufacturing aspects of Galileo. A month later, India joined the Galileo partnership as well. In March 2004, Israel came to an agreement with the European Union on Galileo participation. These partnerships are worrisome to the United States, not only because of the potential for countries such as China to use the applications in ways that potentially threaten the United States, but because of the "technology bleed" among the partners that could result as well. Perhaps most interestingly, in the agreement signed between the European Union and the United States, potential European military use of the system was ignored, though it is certainly likely to be problematic later. Had that issue not been deferred, it is questionable whether a near-term agreement could have been reached.

Bush's New Vision for Space Exploration

In presenting his "Vision for Space Exploration"[32]—manned exploration of the moon, Mars, and beyond—and then not mentioning it again, Bush joined a long line of presidents who seem to forget their enthusiasm for manned spaceflight after leaving the announcement podium. President Ronald Reagan announced multiple space initiatives and then promptly forgot about them. During the development of the Bush vision, shaped by young and knowledgeable space enthusiasts from various Executive Office of the President (EOP) agencies, President Bush was very supportive. He insisted on a broader vision than the "return to the moon" version initially proposed to him and on international participation. Nevertheless, there was no presidential follow-up, except the White House threatening a budget veto in 2005 after Congress cut all programs associated with his vision. Protecting funding can be argued as the only kind of follow-up really needed, but in this case, considering the amount involved, the effort was more symbolic than effectual.

Washington is littered with space policies announced and then forgotten. President George H. W. Bush was actually very committed to a renewed manned space exploration program, but even with his support, the program, called the Space Exploration Initiative (SEI), did not survive his tenure. NASA and manned spaceflight have managed to limp along, but their anemic states will not equate to leadership for very long if the space programs of other countries continue to mature.

Politically, the 2004 Bush space vision was always a vision bordering on fantasy. Though perhaps well intended, it was effectively doomed from the start. The vision as announced was a very broad-brush outline of intent, describing a return manned mission to the moon, as well as manned missions to Mars and beyond. But the devil is in the details, and those details must be in some way attached to reality. Three major circumstantial realities predetermined the outcome of that new vision. First were budget issues. The domestic budget has been, and will likely remain, an effective hostage to the war in Iraq, homeland security concerns, and clean-up for Hurricane Katrina—and like events in the future. Before the budget for fiscal year 2005 even came out, it was expected that there would be no "new money" for any agency other than the Department of Defense (DOD) and the Department of Homeland Security. As it turned out, an exception was made for NASA, which could be—and was—read as a strong short-term program endorsement. The White House requested $1 billion in new money to implement the new space vision, with another $11 billion to come from a reprioritized NASA budget. Relatively speaking, $1 billion is a small amount, especially when technology development is concerned. The Pentagon spends $1 billion daily. Politically, however, it was huge. NASA was the only non-security-related agency to get a budget increase in fiscal year 2005. But the $1 billion was only a drop in the bucket toward program implementation.

Second, and equally critical, the NASA budget was already consumed by commitments to support existing programs. While it is possible to reprogram funds, as would have been necessary to find the $11 billion, all the existing programs have political champions in Congress. The NASA budget has always been among those most strongly tied to pork-barrel funding—to both its benefit and its detriment.

Third, the public view of the NASA program has consistently been that it is desirable, but expendable. The public supports human exploration, and even recognizes that benefits accrue on Earth, but it prioritizes funding for roads, schools, health care, and near-term benefits over space programs, particularly space exploration. Without a motivation beyond exploration, the vision has no legs.

Overall, however, the Bush administration rang the death knell for the program by leaving responsibility for the bulk of the vision funding to future administrations. Washington is driven by politicians looking to make their mark and take credit for achievements. It is politically unimaginable that a future president or presidents will allocate the money necessary to carry Bush's vision of exploration vision forward as a legacy program for a predecessor.

The vision might at least have received token political support had there been any indication of commitment from President Bush. When that did not materialize, members of Congress saw no reason to kill established programs in their districts in favor of a program that the White House appeared not to care about anyway. Members of Congress, particularly in the House of Representatives, are basically in a constant bid for reelection. With a two-year election cycle, they must consistently demonstrate the benefits of their efforts to voters back home, most often and tangibly by jobs and public works projects in their home districts. In the "space states," or states where aerospace industries or NASA field centers are located,[33] politicians covet projects being developed in their areas and have no motivation to forgo commitment to them in favor of an amorphous vision with apparently little backing and a tenuous future. The Kennedy Space Center, for example, is closely wedded to the space shuttle. Any plan to retire the shuttle inherently affects employment in that congressional district. The degree to which representatives protect the jobs in their districts is legendary and long-standing; expecting anything different, without offering replacement jobs or constituent benefits in return, is simply unrealistic.

In July 2004, the House appropriations subcommittee that oversees the NASA budget slashed NASA funding by more than $1 billion—all the "new money" allocated to NASA, and all earmarked for programs related to Bush's space vision. In the committee report, support was expressed for the vision, but not as a NASA top priority. Eventually, after pressure from the White House, $86 million of the $1 billion that had been slashed was reinstated before the appropriations bill was passed. In the meantime, however, NASA found itself with $1.4 billion in additional costs to established programs. These costs had been unanticipated when the original budget request was submitted, and absorbing them cast a large shadow over NASA's ability to undertake anything more than token efforts toward initiating Bush's vision. Continuing problems with the shuttle exacerbated an already untenable situation.

In fiscal year 2006, the NASA portion of the federal budget request was for $16.5 billion, a 2.4 percent increase over the 2005 appropriation; 2007 included a $16.8 billion request for NASA, a 3.2 percent increase over the 2006 budget appropriated for NASA. Both figures represent approximately 0.7 percent of the overall budget. That may be an appropriate amount for exploration, but it is insufficient to address the continuing problems with the shuttle, finish the ISS, and achieve the vision goals within the time line given.

As mentioned earlier, the American public is generally intrigued by space flight and recognizes the benefits, but supports tangible earth-bound programs first. This was the case even during the Apollo program, and continues to be so.[34] That raises the question of why Apollo was able to garner political support to reach fruition in the first place,[35] as other programs have limped along or died along the way.

According to a 2004 Gallup poll, 29 percent of the American public sees the "human nature to explore" as what drives American space exploration. Another 21 percent see maintaining the status of the United States as the international leader in space as the main impetus.[36] The clearly defined rationale for the Bush space vision is exploration. By contrast, Apollo had a strategic purpose, defined as leadership plus national security implications. Implicitly, if not explicitly, Apollo was about "beating the Russians" to the moon, to show that our technology was superior to theirs, and to establish international leadership. There is no such driving motivation associated with Bush's vision.

Hence, the Bush space vision has been doomed to failure from its inception. The follow-on program, announced in September 2005, has been termed "Apollo on steroids,"[37] as its goal is to develop a vehicle to get the United States back to the moon by 2018. It is intended to keep the United States in the manned space business. Failure to keep the United States not only in the manned space business, but leading it, will have broader implications. When the Russians launched *Sputnik* and then Yuri Gagarin, the first man into space, there was a global perception that U.S. technology was seriously lagging. While not factually correct, the perception affected the American strategic position in the Cold War, and the United States had to respond. Today, there is a similar, growing techno-nationalist perception that others—specifically the Chinese—are now "beating" the United States in manned space programs. The Cold War mentality, perpetuated by a "race," real or perceived, aggravates the already unhealthy security dilemma. Therefore, the time is right for another détente effort, to turn competition into cooperation.

Darth Vader: "I Have You Now!"

The commitment of the United States to manned space is tenuous, but its commitment to using space and space technology as key military assets is not. Whether that is a comforting thought or a scary one depends on one's perspective. The United States may see itself as Han Solo or Obi-Wan Kenobi, underdog heroes from the *Star Wars* films, but when the United States speaks of its plans in space, much of the rest of the world hears the eerie voice of Darth Vader, the films' central villain: "I have you now!"[38]

The first Gulf War, dubbed the first space war by General Merrill McPeak of the Air Force,[39] was a first step toward much greater reliance and utilization of military space assets in areas such as intelligence, surveillance, and reconnaissance (ISR), communications, and use of precision-guided munitions, or smart bombs.[40] From Operation Desert Storm in 1991, to Operation Allied Force in Serbia in 1992, to Operation Enduring Freedom in Afghanistan in 2001 and 2002, precision-guided munitions rose as a percentage of total delivered air weapons, from 7.7 percent, to 29.8 percent, to 60.4 percent.[41] Operation Iraqi Freedom continued U.S. reliance on military space assets, with 68 percent of the munitions used in that effort being precision guided.[42] Increased dependence on space assets leads to an increased need to protect those assets. With Air Force Doctrine Document 2-2.1, *Counterspace Operations*, in August 2004, Air Force intentions about protecting U.S. space assets and denying the use of space to potential adversaries were more clearly articulated than ever before in an unclassified document. Intentions include developing and using offensive counterspace capabilities. Counterspace operations are those intended to defend U.S. space assets and capabilities. "Offensive counterspace" basically means "offensive defensive" capabilities: the ability to attack in defense of your own assets, or to deny assets to others. Statements regarding potential space weapon development in the past had always referred to a purely defensive mission. The doctrinal shift puts satellites of all types, including commercial and those from neutral countries, on potential target lists. The boldness of the document also indicates that the Air Force leadership clearly believes that space warfare has the support of civilian leadership. That, coupled with preemption principles embedded in both the 2002 and the 2006 national security strategy, has generated considerable alarm in some countries.

Beyond paper documents, U.S. strategic communications about its intentions in space are conveyed through actions and speeches. A new ground-based system capable of attacking enemy satellite communications, the so-called counter communications system, was announced at an aerospace conference in September 2004 by Brigadier General Larry James of the Air Force, vice commander of the Space and Missile Systems Center in Los Angeles.[43] The United States is vigorously pursuing viable small-satellite technology, even as it has grave concerns about China developing the same technology. An Air Force official, speaking to a reporter from the trade publication *Space News* about an Air Force small-satellite program known as XSS, stated, "XSS-11 can be used as an ASAT weapon."[44] Dropping news items like this, and getting no negative domestic response, allows weaponization to continue incrementally.

The Air Force document states that the United States seeks space superiority, an advantage over other countries by some potentially minimum amount. However, the ultimate American goal appears to be space dominance, the unchallengeable ability to control the space environment. The lineage of this position comes from documents such as *Vision for 2020*, published in 1997 by the U.S. Space Command, which stated, "The emerging synergy of space superiority with land, sea, and air superiority, will lead to full spectrum dominance,"[45] which is defined in Joint Vision 2020 as "the ability of U.S. forces, operating alone or with allies, to defeat any adversary and control any situation across the range of military operations."

The themes of that document were later echoed in a January 2001 congressional report from what was known as the Space Commission. Chaired by Rumsfeld just before he became secretary of defense, the commission warned, "If the United States is to avoid a 'Space Pearl Harbor,' it needs to take seriously the possibility of an attack on U.S. space systems."[46] The commission recommended the creation of the U.S. Space Corps, which would defend our space-based "military capability." In 2003, the Air Force's transformation flight plan[47] included ideas for orbiting weapons to send giant metal rods crashing to earth, officially known as hypervelocity rod bundles, and unofficially called "Rods from God." That document, however, discussed only hardware. The 2004 counterspace operations documents add another component to the trend of developing space as the fourth battlespace, stating when and how such hardware would be used.

While the promulgation of documents describing changes in U.S. space policy has gone generally unnoticed by the mainstream American media and the public, the trend toward weaponization has not gone unnoticed in other countries. When the U.S. government begins publishing documents on Web pages showing lasers firing from space, as the *Vision for 2020* Web site originally did, people and countries tend to get nervous. The "Rods from God," with an artist's rendering provided in the June 2004 issue of *Popular Science*,[48] have generated considerable discussion at scientific conferences, not only about technical viability, but about whom the United States intends to use them against. In 1997, when the United States still embraced multilateralism, countries reassured one another and accepted reassurance from the United States that its intent was benign and defensive. Since September 11, 2001, as the United States has been seen to not only embrace but boast about a primacist grand strategy, employing preemptive tactics, and talking in terms of "preventive war" as the future norm,[49] accepting reassurance about the benign intent of the United States has become increasingly difficult.

The development of space weapons may be necessary and advantageous for the United States sometime in the future. If so, the next question becomes when. Clearly, the United States has the opportunity to explore other options. The consequences of not considering the longer-term challenges of trying to militarily "fix" Iraq, as well as the limitations of relying too heavily on military power alone, have become increasingly clear; the need to think in the long term before embarking on developing space weapons should be apparent.

Exponential Effects

Taken individually, each of the events that occurred between October 2003 and October 2004 described at the beginning of this chapter is interesting, at least to select audiences. Considering them together, however, expands the individual importance and exponentially changes the potential combined impact. While the rest of the world seeks to increase its ability to use space assets for information linkages required for economic growth in a globalized world, the United States sees much of the technology they are seeking as militarily sensitive and, consequently, is trying to stop its spread. That initial clash of ambitions is further exacerbated by the parallel emphasis the United States places on expanding its space superiority to space dominance. The risk of being left behind by the growing technology gap between the United States and other countries, with its clear economic ramifications, carries dire consequences.

Ultimately, the potential for the will of the United States to become absolute is perceived as threatening to the very sovereignty of other countries.

Jervis states that "if one state gains invulnerability by being more powerful than others, the problem will remain because its security provides a base from which it can exploit others. . . . Others who are more vulnerable will grow apprehensive, which will lead them to acquire more arms and will reduce the chances of cooperation."[50] The dual-use nature of space technology lowers the bar for raising the apprehensions of the dominant state—but U.S. apprehensions will not keep other countries from acquiring technology.

In 2004, President A. P. J. Abdul Kalam of India discussed the impact of U.S.-imposed sanctions on India's plans and ability to develop space launch technology, either for peaceful use toward economic development or for missiles. In his view, U.S. sanctions on India were self-defeating. He suggested that technology embargoes certainly delayed and increased the costs of launch technology development in India, but they did not stop it. "No country will be a loser except for the one that imposes the sanctions," he said. "If you don't do business with the country that has had sanctions clamped on it, like in the case of India, it comes out the winner. The guy who got the sanction will win, as it will indigenously develop [what it needs]. This is plain logic. It is only a question of time and money."[51] Today, not only can India launch payloads into space, but it can use and distribute that technology at will, without having to adhere to restrictions that accompany foreign-acquired technology.

When countries feel that their sovereignty may be threatened, they react; reactions range from rhetoric and gestures, to seeking autonomous capabilities, to potentially preparing an asymmetric response to U.S. space capabilities. Examples abound. Besides India developing indigenous nuclear and launch technology, and Europe stepping away from dependence on U.S. space capabilities through Galileo, the European Union is considering autonomous military space programs, previously unimaginable. Japan, a U.S. missile defense partner, has developed the Information Gathering Satellite (IGS) system, its first reconnaissance satellite system. China, feeling the most threatened by U.S. actions, is expanding the development of its own military space capabilities. Space activity by other countries triggers a reaction from the United States, often in the form of further escalation of military space activity. A dangerous action–reaction syndrome results, reminiscent of Cold War efforts, and could well lead to development and early deployment of hair-trigger space weapons.

Although the commitment of the United States to military space is clear, a corollary commitment to manned spaceflight, with its international leadership implications, is not. To date, China has successfully launched two manned launches, twenty-four months apart. When the United States had the political will to send a man to the moon during the Apollo years, it was unbeatable and awesome. Recently, however, the shuttle has been grounded more often than it has flown; the space station is still not complete, though it was announced during the Reagan administration; and the NASA budget is dwarfed by the military space budget. Consequently, regarding manned space, China assumes the role of the tortoise racing against the U.S. hare. China also offers countries partnership options for manned spaceflight outside the United States, and with potentially more favorable partnership terms.

Beyond the prestige that manned spaceflight offers, countries interested in space development but frustrated by U.S. export laws see cooperation with China and other countries, such as Israel, Russia, and France, as a road to developing space technology. The United States has in effect pushed countries toward indigenously developing technology considered potentially threatening to U.S. security efforts. When the United States sells technology, it can control (in the sense of managing) what it sells. When the United States is not involved, it has no control. A zealous interest in military space, including denial of dual-use technology to other countries, especially while lacking a viable and coherent plan for the future of manned spaceflight, is becoming increasingly counterproductive.

The Big Picture

Walter Russell Mead states that there are two categories of power, with two aspects of each.[52] The first category is hard power, the most obvious type of which is military power. The value of space to the military as a dimension of hard power has been demonstrated vividly and repeatedly. There is also what Mead calls sticky power—that is, economic power, to the extent that it can become coercive. Globalization is sticky power on steroids; thus it is a form of hard power. Even the prospect of Galileo making GPS commercially obsolete was enough to prompt the United States to negotiate with the Europeans. The second category, referencing Joseph Nye, is soft power.[53]

Soft power is about reaching goals through popularity and persuasion rather than coercion. People or countries work with others because they like them or what they stand for. Efforts to build, or even sustain, that area of U.S. power were noticeably lacking during and immediately after OIF. Nye spoke on soft power at a September 2003 conference of military leaders in Washington, followed on the podium later in the day by Rumsfeld. When Rumsfeld was asked about soft power, he said he did not know what the term meant,[54] reflecting the priority that the Bush administration gave it. At least regarding Iraq and the global war on terrorism, attitudes have since changed. Bush's close friend and adviser Karen Hughes was sworn in as the State Department's undersecretary for public diplomacy and public affairs, and ambassador for the same, in September 2005; she was charged with transforming the image of the United States, a component of soft power. The Pentagon instituted its Countering Ideological Support for Terrorism (CIST) campaign to learn more about foreign cultures as part of its effort to "win the hearts and minds" of potential radicals. Clearly, not all battles can be won with guns.

Soft power, according to Mead, can be "sweet" or "hegemonic." Sweet power refers to countries sharing cultures and values, or a country admiring and seeking to emulate the culture and values of another. Hegemonic power is an interplay of the other three types of power, by which siding with the hegemon is determined to be better than any other option. Countries may see U.S. actions as overbearing in the global war on terrorism—including the war in Iraq—but few Westerners at least would choose to live under the Taliban instead. The Apollo space program provided the United States with both sweet and hegemonic power, but, unfortunately, those aspects of foreign policy generally, and as related to space specifically, are too often ignored. That is not in the best interests of the United States. Soft power is key for accomplishing tasks essential to the continued effective use of space by the United States.

Comparing the United States and Great Britain as the keepers of a stable world, Mead points out that while the United Kingdom has full use of all types of power, hard and soft, the United States is more limited. The Iraqi government under Saddam Hussein could be toppled by U.S. military (hard) power alone, but nation-building and keeping the country from imploding has proved more troublesome. Regardless of desire, it is questionable whether the United States has the military personnel to move from place to place on the globe dismantling undesirable regimes. While fighting terrorists, instilling fear of the United States may be necessary in some instances. Nevertheless, rebuilding Iraq has demonstrated that, at some point, soft power becomes imperative to making and maintaining peace. In space, cooperation is imperative for tasks such as allocating orbital satellite slots, broadcast frequencies, and debris management, all critical for the United States to maintain its lead.

Historically, cooperation and competition in space have been hallmarks of U.S. strategy. The current U.S. view of space activity is primarily competitive and based on hard power. However, the technology genie is out of the bottle and cannot be put back. Other countries will obtain space technology, and with increasingly fewer controls because of the expansion of autonomous industrial sector capabilities. Further, other countries—even potential adversaries—need not seek technological parity to challenge U.S. assets. They need only seek those technologies they want to utilize themselves and employ asymmetrical approaches to thwart U.S. advantages. In missile defense, countermeasures—actions or hardware designed to confuse or overwhelm—are much easier and cheaper to develop and deploy than is missile defense technology itself.[55] Therefore, relying on only the hard power aspects of space, usually involving technology, is inherently short-lived and, ultimately, counterproductive. It places the United States on a constant quest to create unassailable hardware—an expensive, if not impossible, quest. The United States would be better served using all the diplomatic, informational, military, and economic tools at its disposal to protect its strategic space assets.

The United States will likely find itself unable to prevent the technological expansion that is a fact of life in a globalized world. Giving other countries a vested interest in protecting space assets—through ownership or dependence on use—leverages global needs rather than trying to quash them. Offering opportunities for countries to work together on projects of mutual benefit and interest, under U.S. leadership, can be to the advantage of the United States.

The United States has historically used manned spaceflight to demonstrate its international leadership and build soft power. Today, we are implicitly ceding our leadership. During the Cold War, the Apollo program demonstrated U.S. technical prowess for the soft power that would accrue to it. Approval for maintaining the funding levels needed to sustain Apollo through completion were possible (and yet just barely) because to stumble would have been a strategic failure, unacceptable in the Cold War, even though the program was originally another administration's idea. Enhancing U.S. soft power through a cooperative, international manned space exploration program, especially in contrast to the threat perceived from U.S. military space dominance as a hard-power advantage, is a strategic goal that the United States has neglected at its peril.

The United States must also ratchet down the security dilemma that currently characterizes space activity. That means reining in its space control and force application programs. To do so, the United States will need assurances that its own space assets are secure. Other countries might be more willing to consider those assurances if their own fears of being exploited were diminished. Yet there are no efforts currently being made toward those goals; in fact, quite the opposite is occurring. Jervis generalized about the reasons options are overlooked:

> Statesmen who do not understand the security dilemma will think that the money spent is the only cost of building up their arms. This belief removes one important restraint on arms spending. Furthermore, it is also likely to lead states to set their security requirements too high. Since they do not understand that trying to increase one's security can actually decrease it, they will overestimate the amount of security that is attainable; they will think that when in doubt they can "play it safe" by increasing their arms. . . . The belief that an increase in military strength always leads to an increase in security is often linked to the belief that the only route to security is through military strength. As a consequence, a whole range of ameliorative policies will be downgraded. Decision makers who do not believe that adopting a more conciliatory posture, meeting other's legitimate grievances, or developing mutual gains from cooperation can increase their state's security, will not devote much attention to these possibilities.[56]

The stakes are too high not to consider all options. Both developed and developing countries are increasingly becoming interested in creating space programs of their own, and are likely to do so with or without the cooperation of the United States. To retain its competitive edge, the United States must step back from its stance of seeking to stop the development and spread of hardware, software, and knowledge that could be used in space programs. This desire to monopolize space technology is perhaps most visible in U.S. regulations of dual-use technology, which is where our discussion begins.

TWO

The Conundrum of Dual-Use Technology

A lot of the things that Iraq asked for [for reconstruction] were not allowed because they came under what is called dual use. For example, at some point, it was ridiculous enough that pencils for school students were not permitted because they contain lead, and lead could be used as a dual thing.

—Youssef Ibrahim

A scientist from the National Weather Service tells a story of going before a congressional committee to request funding for a new weather satellite. He explained what new capabilities the satellite would offer, who it would serve, and what it would replace. When he finished, the committee chairman, in a rumpled suit and country-boy accent, leaned forward and replied in a tone dripping with misgiving, "Now sir, I know you and your scientist friends would like a new satellite—and I'd like to be able to give it to you—but I really can't justify it. After all, we already have the Weather Channel."

Few people realize how much the effects of space and space programs touch their lives on a daily basis. The civilian benefits of space, however, have gone far beyond the Tang and Velcro created through the Apollo program. Whether through space services directly, including weather satellites and communication satellites, or technology developed for space systems and used on earth, space and space-related technology is increasingly a part of everyday life. Software developed to target precision-guided munitions in battle is basically the same as that used to guide laser beams in minimally invasive surgery that allows patients to walk back to their hospital rooms after brain or heart operations. The value of technology useful to both the military and civilian-commercial sectors is enormous. But that same technology also presents problems.

Because of the military value of dual-use technology, governments consider much of it to be "sensitive" and therefore subject to government controls when sold for export. Most governments, especially in developed countries, have controls of some type, but U.S. export-control laws are considered to be the most draconian by far. They are both restrictive regarding what is controlled and ambiguous regarding how such control is wielded. Part of the problem with writing export-control laws on dual-use technology is that while the intent of trying to control technology that is potentially beneficial to other militaries is good, the reality of trying to do so has become progressively more absurd. The increasing commercial availability of dual-use goods extensively complicates considerations about what should be subject to control. Technically, the computing powers of some Apple computers and the Sony PlayStation 2 game console are high enough that they could be classified as "sensitive" military technology, and therefore subject to export-control laws designed for weapons. Night goggles, sold on the Internet by a multitude of suppliers and at one time through the L. L. Bean catalog, have been considered munitions by the State Department and therefore theoretically subject to government control. Whether such control can be realistically exercised is another matter.

According to the Department of Defense, dual-use technology is a technology that has both military utility and sufficient commercial potential to support a viable industrial base.[1] Between 1997 and 2000, the DOD and various private industries invested more than $800 million in over 300 projects to develop dual-use technology, a clear indication that the government considers this technology to be militarily and commercially advantageous. One of the projects involved developing an affordable, highly reliable antenna, suitable for both weapons system delivery and cellular communications. Dual-use technology provides potential use and profits to multiple markets.

The DOD's interest in the development of dual-use technology began in earnest after the Cold War. By then, the U.S. space industry had evolved into four distinct sectors: military, intelligence, civilian, and commercial space. Since 2001, the Air Force has been the "executive agent" for the military space sector, though all branches of the service are heavy space users. Intelligence sector assets, sometimes referred to as "national systems," are most often associated with the National Reconnaissance Office (NRO), whose very existence was classified until 1992. Meanwhile, NASA has represented civilian space since the organization's inception in 1958. Historically, except for communication satellites, the commercial space sector was dependent on the plans of the other sectors for business. It provided hardware, rather than being an operator and a provider of services itself. That, however, has changed dramatically in recent years.

Since World War II, it has been recognized both in the Pentagon and by potential adversaries that U.S. military superiority is directly linked to its technical advantage. It has been recognized only more recently in the Pentagon that new technologies critical to defense are emerging in the commercial sector. Information technology fuels the engine of globalization and makes a great deal of money. Market revenue in telecommunications grew from $523 billion in 1991 to $1.37 trillion in 2003. During the same period, telecommunication services capital expenditures rose from $124 billion to $215 billion.[2] The profit potential in telecommunications is high, and it represents only one information-technology sector. The private sector is investing in information-technology research and development (R&D) at rates that the government cannot keep up with, and will continue to do so as long as profit returns are anticipated to warrant those investments.

The commercialization of technological development has four consequences. First, the government does not have to spend valuable R&D dollars to duplicate what the private sector is already doing. When the government wants to offer incentives for R&D in a certain direction, or feels the imperative to maintain an in-house capability, it can pick and choose these situations, allowing government R&D investment dollars to go further. Second, the U.S. government must work harder to ensure that it has early access to these technologies. The establishment of the Dual Use Science and Technology (DUST) program in the Pentagon evidences concerted government efforts to foster innovation with potential value in a manner that ensures the government access. That program's stated purpose is "to partner with industry to jointly fund the development of dual use technologies needed to maintain our technological superiority on the battlefield and for industry to remain competitive in the marketplace."[3] As the government finds itself at the mercy of supply and demand, it must be a smart consumer. Third, the government, including the military, becomes increasingly dependent on commercial technologies. During Operation Iraqi Freedom, there was a more than 560 percent increase in the use of commercial satellites (comsats) for military communications over the pre-OIF period.[4] Commercial comsats supply about 80 percent of the U.S. military capability for both communications and imagery transmission. Finally, the U.S. government cannot assume that it can prohibit other potential "customers" from buying the same technology that it is interested in. Saddam Hussein's government purchased Russian-made GPS jamming equipment on the Internet during OIF. Globalization and globalized industries are two-sided coins—or double-edged swords, as the case may be.

A globalized world, with the United States as a major proponent, promotes linkages and the rapid expansion of free markets. Subsequently, exports are increasingly assumed to be part of any sales strategy for manufactured goods. This creates issues when dual-use technology is involved. The basic question becomes how to control the spread of militarily useful technology, while encouraging the spread of technologies for economic growth and development. Because almost all space technology is dual use, the challenge regarding space assets is titanic.

Space Technology

According to the Pentagon's Militarily Critical Technologies List (MCTL), "Recent studies have shown that approximately 95 percent of space technologies can be categorized as dual use" (table 2.1).[5] The close link between information technology and space accounts for much of that percentage. For the military, space represents "the high ground," a place to put hardware from which to gain military advantage. For others, space hardware is an essential part of globalization. Space is a critical part of the global information infrastructure, a term coined during the Clinton administration for the multiple networks comprising what was then commonly referred to as the "information highway." Information is inherently valuable to the civilian sector as well as to the military sector, so dual-use space technology abounds.

Information technologies, particularly space technologies, were once limited to wealthy countries and, in many cases, to only the national security sectors of those countries. Today, as the price of acquiring technology falls, they are rapidly becoming affordable to all countries and even individuals. Capabilities that the DOD spent billions to develop in the 1980s are increasingly available to others at a fraction of the cost,[6] through buying either services or the hardware itself. Remotely sensed imagery taken from space, imagery potentially accurate enough to allow identification of significant detail, was once the purview of only select national security communities. That is no longer the case. The NASA Landsat program, started in the 1970s, collected unclassified imagery. In that case, the imagery was electro-optical—basically a camera. Other types of imagery include radar and infrared. Initially, the most precise resolution for a satellite image available on Landsat was 30 meters, meaning that an object had to be at least 30 meters in size to be observable. In 2000, resolution improved to 15 meters. Landsat imagery covers large swaths of land 185 kilometers across, making it useful for such tasks as land management, environmental monitoring, and urban planning. In 1985, it became part of the first major effort made by the United States for the commercialization

Table 2.1 Examples of Space-Related Dual-Use Technology

Generic Technology Application	Civil Use	Military Use	Dual Use?
Communication satellites	Fixed satellite service	Fixed satellite service	Yes
	Broadcasting satellite service	Broadcasting satellite service	Yes
	Mobile satellite service	Mobile satellite service	Yes
	Land, maritime, and aeronautical mobile intersatellite services	Land, maritime, and aeronautical mobile intersatellite services	Yes
Satellite remote sensing/imaging	Earth resource observations	Reconaissance systems	Yes
	Environmental monitoring	Environmental monitoring	Yes
	Meteorological services	Meteorological services	Yes
	Atmospheric research		No
	Geophysics and geodesy	Precision targeting	Yes
	Oceanography	Oceanography	Yes
	Cartography	Cartography	Yes
		Nuclear test detection and surveillance	No
		Early-warning system	No
	Treaty compliance verification	Treaty compliance verification	Yes
Satellite navigation systems	Navigation services	Navigation services	Yes
Rocket propulsion	Space launch systems	Space launch systems	Yes
	Space propulsion systems	Space propulsion systems	Yes
	Sounding rockets	Sounding rockets	Yes
	Orbital reaction control systems	Orbital reaction control systems	Yes
		Ballistic missiles	No
		Antiballistic missiles	No
		Antisatellite systems	No
		Interceptor propulsion	No
		Escape/evasion systems	No

(continued on next page)

Table 2.1 Examples of Space-Related Dual-Use Technology *(continued)*

Generic Technology Application	Civil Use	Military Use	Dual Use?
Satellite search and rescue systems	Search and rescue services	Search and rescue services	Yes
Space science and exploration	Astronomical research		No
	Astrophysics research		No
	Comet research		No
	Space sciences and microgravity research		No
	Exploration	Exploration	Yes
	Planet research		No
	Solar physics research		No
	Environmental definition	Environmental definition	Yes
Technology demos	Technology demos	Technology demos	Yes
Telemetry, tracking, and control systems	Telemetry, tracking, and control systems	Telemetry, tracking, and control systems	Yes
	Monitoring and data collection	Monitoring and data collection	Yes
Permanent manned orbiting stations	Observation laboratories	Observation laboratories	Yes
	Mission staging	Mission staging	Yes

Note: It is interesting to note that since dual uses originally were developed and cited, it could easily be argued that even more have been identified. For example, whereas solar physics research is identified as not being dual use, solar physics as defined today, including the term "space weather," is of significant concern and interest to the military because of the sun's potential damaging impact on military satellites. Dual use also does not differentiate between possible and optimal use of technology. For example, whereas manned space stations technically can be used for observation laboratories, whether there is an advantage to having a manned over a robotic capability is left unaddressed.

Source: Based on a table developed by Stephen E. Doyle, *Civil Space Systems: Implications for International Security* (Aldershot, Eng.: Dartmouth, 1994), 4, and cited in A. Houston and M. Rycroft, *Keys to Space: An Interdisciplinary Approach to Space Studies* (Boston: McGraw-Hill, 1999), 2–8.

of space. The Earth Observation Satellite Company (EOSAT)[7] was selected to commercially operate the Landsat system, allowing customers to buy Landsat imagery by "scene."

With newer commercial satellites, customers can task the satellite, custom-ordering optical imagery of a particular place at a particular time. Custom panchromatic imagery of approximately 1 meter in resolution became commercially available in 2000 from the *Ikonos* satellite, currently owned and operated by GeoEye.[8] Imagery of that precision can potentially be used for a variety of purposes, including military purposes. GeoEye is not the only company offering high-resolution imagery.[9] In March 2002, another commercial company, DigitalGlobe, began offering 60-centimeter resolution imagery from its *Quickbird* satellite, launched the previous October. Imagery below 50 centimeters will be available in 2007. Free high-resolution imagery is available on the Internet through Google Earth, a piece of software that combines satellite and aerial images with mapping capacities. The images are of such quality that several countries, including India, South Korea, and Thailand, expressed concern in 2005 regarding its availability.[10] Is this availability good or bad? The question is really moot, because the technology is already available.

The United Nations has used commercial imagery to detect and inventory the cultivation of poppy and coca plants. Commercial imagery has also been used to improve airplane navigation though narrow mountain passes in Alaska. Certainly both of those examples are highly beneficial. In 2001, when the downed U.S. EP-3 aircraft was sitting on the tarmac on Hainan Island in the hands of the Chinese, near real-time images of the airplane appeared on the Space Imaging Web site. Some analysts argue that the imagery was useful simply because of the transparency that the images provided. Others suggest that such transparency was potentially complicating, if not dangerous, while tense negotiations for returning the crew were being conducted. The dilemmas of dual-use technology extend far beyond the proliferation of hardware concerns and into real-life usage issues.

Some examples of dual-use space technology are easy to identify. Rockets, launchers, and missiles for military use rely on the same basic parameters as space flight for success. If a country can build a missile, it clearly has the technical knowledge to build a launch vehicle. The primary technical differences between the two are trajectory, payload, and guidance systems; the primary difference overall is intent of use. Images taken by remote-sensing satellites can be used to maximize crop rotations for increased yield or to target weapons. Navigation satellites, such as those in the GPS, keep civilian airliners properly spaced and on course as well as guide munitions with precision accuracy.

Other technologies are more difficult to identify as potentially dual use. In negotiations with the United States in the 1970s on a potential antisatellite weapons treaty, the Soviet Union included the space shuttle on its list of what it considered U.S. ASAT hardware. The Soviets argued that the shuttle could approach a satellite in orbit, pluck it out of the sky with its robotic arm, and either destroy it or bring it back to earth. The United States saw the shuttle differently. The example points out the dilemmas posed by perceptions of possible intent.

Component parts are perhaps even more problematic than large pieces of hardware in determining intent. Military and commercial space systems share components for electronics and computers, optics, propulsion, and sensors. Even if the end-use intent is in a commercial system, there is a view, expressed in the MCTL, that "commercial communications satellites will in many cases drive the capabilities of many of the space technologies, which will then be used in all military space systems."[11] If that view is persuasive, then "control" becomes interpreted in the strictest sense and is often a euphemism for "stop." Briefly examining four areas of dual-use space technology more clearly illustrates the complexity of the issues.

Communication Satellites

A comsat is a spacecraft that orbits the earth relaying radio signals. Signals are transmitted from stations on the ground, called earth stations, to satellite transponders—the circuits that receive, modulate, and amplify uplinked signals—for relay to other earth stations. Comsat use in daily life includes a multitude of functions, such as providing direct-to-home television, tracking packages, videoconferencing, sending faxes and pages, supporting high-speed Internet use, and making mobile and long-distance phone calls. As already pointed out, IT is a high-growth global industry. In the United States, cell-phone subscriptions rose from 86 million in 1999 to 182 million in 2004, with cell phones composing about half of the telephone service provided. Chinese cell-phone subscriptions increased from 43 million in 1999 to 335 million in 2004, with cell phones accounting for almost 100 percent of service in China.[12] In short, service provision from comsats is big business.

As of 2004, an estimated 880 satellites were in orbit, including 443 commercial communication satellites and 99 military communication satellites. Of those, 234 are commercial communication satellites in geostationary earth orbit (GEO).[13] For technical reasons, different satellites function in orbits at different altitudes. A bit of orbitology is in order, albeit at a technical level suitable for poets, to understand why certain orbits are useful for certain satellite functions.

Most spacecraft today fly in one of three basic orbits:[14] low earth orbit (LEO), at an altitude of less than 2,000 kilometers above the earth's surface; medium earth orbit (MEO), at an altitude of about 10,000 kilometers; and GEO, at an altitude of approximately 35,800 kilometers. The space shuttle flies in LEO, as does the ISS. Because the LEO is the closest to the earth, it is also used for many imaging satellites. The closest view affords the best picture. The MEO is an especially harsh space environment because the Van Allen radiation belts—swirls of radioactive particles potentially damaging to solar cells, integrated circuits, sensors, and humans—are located there. The MEO is home to the GPS satellites. There are also mobile comsats in the LEO and MEO. The GEO is a unique orbit, in that a satellite in GEO rotates at a speed so as to remain hovering over a specific spot on the earth's surface. The distance from earth is also such that three equally spaced satellites in GEO can provide almost full earth coverage, especially important for providing or relaying communication transmissions. Because satellites in GEO hover over one spot, ground antennas critical to signal reception do not have to move. Subsequently, most of the large communication satellites are in GEO.

There are two major problems currently associated with comsats: orbital crowding—that is, running out of room—in GEO, with increasingly potential for radio frequency (RF) spectrum interference, and running out of RF spectrum. Regarding crowding, satellites must be placed certain distances apart to avoid broadcast signal overlap. Although those distances are getting smaller due to technological improvements, there is still a finite amount of space in GEO. GEO satellite "slots" are an increasingly scarce commodity and a source of competition among both countries and organizations within countries.

The RF spectrum—that part of the electromagnetic spectrum in which radio signals are received and then rebroadcast from satellites—is also a limited natural resource. There is only so much spectrum available to satisfy the high demand from users and potential users. The RF spectrum is divided into portions called frequency bands. These bands are measured in hertz, and the wider the band, the more information can be transmitted—hence the term "broadband" being associated with faster Internet connections. Since the use of radar in World War II, these bands have been commonly designated by letters. Commercial operators providing video and data services typically use the C-band (4–8 GHz) and the Ku-band (12–18 GHz). The S-band (2–4 GHz) and the L-band (1–2 GHz) are used for mobile phones, ship communications, and messaging. Broadband communications utilize the Ka-band (27–40 GHz). Many U.S. military satellites use the K-band (18–27 GHz) and the X-band (8–12 GHz).

However, as mentioned earlier, demand for spectrum allocation is high and growing, and there is little unallocated space for new users in the usable range. The demand for more access or more bandwidth by militaries, particularly the U.S. military, is nearly insatiable. Militaries use bandwidth for all the communications tasks that civilians do, and more, including command and control of weapons systems. Admiral James Ellis, as commander of the U.S. Strategic Command, was quoted as stating about bandwidth demand, "We've got no clear view of when it might slow down. I believe that it will almost certainly never stop—the only question is the pace at which it continues to grow."[15]

The organization in charge of allocating GEO slots and international coordination of the RF spectrum is the International Telecommunications Union (ITU) in Geneva, Switzerland. Countries manage their use of the RF spectrum domestically, but they must do so consistent with ITU allocations, which include international interference coordination. The increasing demand is due to the IT explosion that both fostered and facilitates the latest round of globalization. Subsequently, the equitable sharing of orbital slots and RF spectrum, as facilitators of globalization, is a major concern for both developing and developed countries and, consequently, the ITU.

The tension between civilian and military demand for capabilities takes an interesting twist in the case of comsats, as militaries often rely on commercial comsats for their own use, as illustrated by usage numbers during OIF. It can be argued that OIF is an extraordinary case and not the norm, but it is expected that by 2010, commercial firms will provide 60 percent of the military's communication services and 30 percent of its remote-sensing services. As the single largest customer for the satellite industry, the DOD generated close to $500 million in commercial communication satellite revenues in 2004, and the demand is expected to double by 2010. In November 2004, the Pentagon announced that it intended to launch $18 billion worth of special-purpose global communication satellites for its own use, but those will not alleviate the need for additional commercial satellite use. Assets such as unmanned aerial vehicles, which are expected increasingly to be used in the future, are operated through commercial satellites.

The opportunities and dilemmas created by the economic, political, and military values of communication satellites can only be expected to increase in the future. The U.S. government has already declared communication satellites, for export purposes, as sensitive technology subject to the same export controls as are exerted over weaponry. To those who see comsats as globalization tools, the control of comsat technology can smack seriously of an effort to deny technology critical for development to other countries, perhaps deliberately. Not to worry, though—other countries are eagerly and quickly stepping in to sell comsats to prospective customers outside the United States, to the detriment of the American satellite industry, on which the military relies. That situation discussed further in chapter 6.

Observation Satellites

In the early days of the space race between the United States and the Soviet Union, before the launch of *Sputnik* when the race went public and goals then shifted, each side had a different priority, based on national security needs. The Soviet Union sought a space launcher (a missile) to use as a long-range nuclear weapons delivery system. The United States already had long-range bombers capable of reaching Soviet targets, so it was more interested in developing satellites to look behind the Iron Curtain into the closed Soviet Union. That too, however, would require a launch system. Much of the technology development occurred simultaneously. Surveillance was not as high a priority for the Soviet Union, because information about the United States, as an open society with a free press, was far more readily available. Before satellite technology was available, the need for information about the Soviet Union drove the United States to use high-altitude aircraft, such as the U-2. Although the United States denied that the flights were taking place, the program met with disastrous consequences when a spy plane piloted by Francis Gary Powers was shot down over the Soviet Union in 1960.

After fifteen years of scientific and technical effort, in August 1960, the United States had its first surveillance satellite success, with the *Discoverer XIII* satellite. The program was actually a cover for Project Corona, a spy satellite program approved by President Dwight D. Eisenhower in 1958. These early satellites sent their crude optical imagery back to earth through film canisters that were ejected into the atmosphere, and then retrieved in midair while parachuting down. Real breakthroughs in the technology began with the Keyhole program, first launched in 1959. The Soviet Union launched its first surveillance satellite, *Kosmos IV*, in 1962. By that time, the military potential of the new technology was clear. That same year, President Kennedy created an interagency committee to study the political ramifications of spy satellites and the way to facilitate civilian needs while limiting access to high-resolution data, resulting in a separate system specifically for civilian use. When initially created in 1967, the civilian system was called the Earth Resources Technology Satellite Program (ERTS) and later became Landsat.[16]

Landsat was the first remote-sensing program to offer imaging data commercially. Having a monopoly on the market allowed it to grow somewhat complacent, until technology improvements were spurred by the launch in 1986 of the competitive French *Satellite pour l'observation de la terre* (*SPOT*), which offered 10-meter resolution data. The Russians joined the commercial market soon thereafter, offering 5-meter imagery from their former military satellite systems, the KFA-1000 and the MK-4. The commercial market had arrived.

The commercial imagery market evolved throughout the 1990s and caused the military great concern. By the early 1990s, Russia was selling declassified imagery with 2-meter resolution, rendering 30- or even 10-meter Landsat data simply not competitive for many uses. As commercial ventures began to infringe on two areas of formerly exclusive military purview, the military became worried about the increasingly higher-resolution data being offered and the timely acquisition of that data. But the United States did not have a monopoly on imagery. Thus U.S. policy either had to allow U.S. industry to compete, or it would be left behind.

In 1992, the United States passed the Land Remote Sensing Policy Act, which included provisions for licensing—that is, controlling—commercial earth-observation satellites. In the last days of the George H. W. Bush administration in January 1993, the first license was issued to WorldView Imaging Corporation in Livermore, California, for a 3-meter resolution satellite. In March 1994, President Clinton issued Presidential Decision Directive (PDD) 23, setting down the specific regulatory guidelines for U.S. commercial remote-sensing ventures. These licenses and regulatory guidelines allow the sale of high-resolution satellite imagery from American firms to domestic and foreign buyers.

Commercial uses of satellite imagery are varied, and are often combined with other technologies as well (such as the GPS) for even broader and more complex applications. The availability of this imagery has served global purposes, from peaceful settlement of border disputes and effective management of sprawling refugee camps to verification of arms control agreements.[17] In 1996, satellite imagery was used to convict those responsible for an oil spill in Singapore harbor. With many ships in the harbor, all had plausible deniability, until an image taken perchance by a European Space Agency satellite was discovered and made available. With that, the source of the oil leak became obvious, and the ship owners were held responsible in court.[18]

The value of commercial satellite imagery in the civilian market is clear. But there is value to militaries as well. Commercial companies do substantial business with the U.S. military and the militaries of other countries. The benefits offered to the civilian sector therefore could, during times of conflict, also serve hostile forces or individuals, potentially negating the advantages of friendly forces. It is with these possibilities in mind that the Air Force's doctrine, outlined in *Counterspace Operations*, talks about being duty-bound, should the need arise, to negate space assets that could aid the enemy, without that asset having to belong to the enemy. According to the doctrine, the asset could belong to a country or company with no direct role in the war. That kind of action, however, could raise thorny issues associated with respecting neutrality under the Law of Armed Conflict.[19]

So far the discussion has focused on optical imagery, but radar and infrared imagery present unique issues as well. For the military, radar imagery provides the capability to "see" through cloud cover, which optical satellites cannot do. Further, it provides not just an image, but depth perception, which can be critical when trying to distinguish topography. What appears in an optical image to be a hedge running across a field could actually be a large, impassable ditch. Radar satellites can avoid these misinterpretations.

Infrared imagery satellites provide "heat signatures." Sensors on board look for heat, found in conjunction with missile launches and explosions. At a much lower level, they can also detect body heat. Infrared capabilities gained notoriety in the 1992 movie *Patriot Games*, based on the Tom Clancy thriller. In that movie, British commandos attack a terrorist training camp, which the hero, Jack Ryan (Harrison Ford), supposedly watches through a sophisticated infrared imagery satellite that relays the action like a television image. Unfortunately, liberties are taken in the scene with both physics and technology, so a realistic portrayal of capabilities ends up being sorely lacking in the movie.[20] Satellite sensors, however, can provide a variety of valuable capabilities.

Navigation Satellites

Developing military hardware is accomplished by funding specific programs. Program development, including for space programs, is supposed to be driven by identifying requirements or needs. After a mission is identified, a capability is developed to successfully accomplish that mission. The practice was implemented to foster good stewardship of federal funds and is part of a time-tested approach to force planning. It is intended to avoid developing technology simply for the sake of development. It can, however, have its drawbacks. Sometimes the "need" is not yet known, or it is unclear. Sometimes a good idea cannot be matched to a specific mission in advance. In 1979, the Office of the Secretary of Defense (OSD) cut the development budget for a fledgling GPS program by $500 million, or 30 percent of its total budget, between fiscal years 1981 and 1986. The reasoning was simple: while supporting the concept, the military services could not identify specific missions, or enough specific missions, to justify developing the system.[21] The OSD knew what it would do with a tank or a gun, but not necessarily the GPS. However, while OSD cut program funding, it did not cut as much as military services recommended. In that regard, OSD was instrumental in keeping the GPS alive; it put the GPS back in the budget when the services zeroed it out between 1980 and 1982. Now, of course, the military is heavily dependent on the GPS.

The GPS is a constellation of twenty-four satellites (plus spares) that orbit the earth every twelve hours at an altitude of 14,000 miles from the earth's center. The system is owned and operated by the DOD, specifically the Air Force. Each satellite carries four highly accurate atomic clocks that "tick" to an accuracy of 1 nanosecond (one-billionth of a second); future satellites are expected to be accurate to between .01 and .1 nanosecond. The satellites broadcast their precisely timed radio signals through the atmosphere and onto the earth's surface at the speed of light. The signals from each satellite arrive at any particular point on or above the earth's surface at slightly different times. GPS receivers passively calculate where they are and what time it is by comparing the signals of multiple satellites in the constellation. Receivers require an unobstructed view of the sky, so they can be used only outdoors, and they often do not perform well in forested areas or near tall buildings.

The degree of accuracy of GPS signals varies. Initially, timing errors were deliberately inserted into the transmissions to downgrade the accuracy of nonmilitary GPS receivers. This practice, called selective availability (SA), was intended to discourage foreign military exploitation, but the argument was overtaken by another technology, differential GPS (DGPS), which could circumvent SA through an additional receiver fixed at a known location nearby.[22] Observations made by the stationary receiver correct positions recorded by the mobile units, producing an accuracy of about 1 meter. The SA "switch" was subsequently turned off by President Clinton in May 2000. Now, the accuracy of a GPS-determined position depends on the type of receiver being used. Most handheld GPS units, widely available in retail stores and on the Internet for upward of $100, have about 10-meter accuracy.[23] Military receivers, in addition, use the encrypted so-called P-code—precise or precision code—to increase accuracy.

The importance of precision for GPS, especially when used with other satellite systems, cannot be overstated. Responsibility for the GPS satellite constellation belongs to the Second Space Operations Squadron at Schriever Air Force Base, Colorado. In 1996, a timing error was transmitted from the GPS control center to one GPS satellite for six seconds before being caught by an alert operator. Over 100 commercial cell-phone networks on the East Coast were affected, some for up to twenty-four hours. This illustrates how important the program has become in a very short time, despite having no clearly defined mission at conception, and its significant financial implications. It also explains why the GPS has become known as one of the first global utilities.

The experience of just one company shows how lucrative GPS-related business has become. In 2003, *Forbes* added Garmin International co-founders Gary Burrell and Min Gao to its list of the 400 Richest People. Burrell and Gao founded a company, initially called ProNav, in 1989. In 1990, they introduced the GPS 100 Personal Navigator, a $2,500 boating navigation aid. The company's first customer was the U.S. Army during the Gulf War in 1991—another illustration of the benefit to the military of dual-use technology developed in the private sector. In 2003, after the company had expanded its product line by adding a series of personal GPS products, such as those designed for recreational runners, and integrated PDA–GPS receivers, the market value of the company stock was estimated at $4.5 billion, with $465 million in annual revenue and $142 million in profits.[24] It should be no surprise that investors worldwide are interested in this technological gold mine.

For the military, joint direct attack munitions (JDAMs) are another example of how GPS has grown in importance. A JDAM is a low-cost guidance kit that can be attached to previously unguided free-fall bombs, converting them into accurately guided "smart" weapons. The kit consists of a new tail section that contains an inertial navigation system and GPS. The DOD began purchasing these relatively low-cost kits in 1998. In OIF, 6,542 smart bombs were used between March and April 2003, composing about 35 percent of the munitions dropped.[25] This capability alone makes GPS a valuable force enhancer. Adding other capabilities—from tracking and identifying friendly and unfriendly forces in the desert, to using in aircraft for precision location identification,[26] to minesweeping operations and more—makes it indispensable.

Therein lays the rub. Not only must the U.S. military protect the GPS for its own use, but it must try to inhibit other countries from using the navigation satellites against the United States—not an easy task, as the GPS, through its extensive and expanding civilian use, is already considered a global utility.

Launch Technology

The history of launch vehicle development illustrates how U.S. fears of losing its space launch hegemony, and the actions it has taken because of those fears, resulted in losing not only its space launch hegemony, but the majority share of the commercial launch market as well. History seems now poised to repeat this *Bonfire of the Vanities* episode in other areas. While officials in the U.S. government were not surprised by the Soviet launch of *Sputnik* on October 4, 1957, they did not anticipate its resultant psychological shock to the American public.[27]

Several factors affected the timing of the first U.S. satellite launch, which came in second to the Soviets. First, the 1954 government-sponsored Killian Report, named for its chair, James Killian, then president of Massachusetts Institute of Technology, had given its highest recommendations to developing an intercontinental ballistic missile (ICBM), along with increasing satellite intelligence capabilities. The Air Force immediately made developing the Atlas missile a priority program. The Army and the Navy were to work on an intermediate-range ballistic missile (IRBM) together, based on a modified Redstone launcher that the Army already had in development, known as Jupiter. Because of the unique technical needs of missiles to be fired from ships, the Navy immediately received permission to pull out and work on its own program, called Polaris. All the services wanted to develop launchers, with the steady flow of funding that such work brought.

The same year, Soviet chief designer Sergei Korolev received approval to develop an ICBM. The Soviets also indicated in 1954 that they intended to put a satellite in orbit as part of the International Geophysical Year (IGY), which extended from July 1957 to December 1958. The IGY, initiated by the international science community, was an umbrella under which numerous countries undertook scientific programs as a symbol of the universality of science, in contrast to the nationalistic Cold War. Because most of the science programs were nationally sponsored, however, there were clearly competitive undertones.

In 1955, the ICBM development race intensified. The Air Force got permission to build a second ICBM, called Titan, and also jumped into the IRBM sweepstakes with a program called Thor. Subsequently, the government was concerned that the IGY efforts would interfere with defense efforts. Hence, National Security Council directive NSC 5520, called the U.S. Scientific Satellite Program, was secretly issued, stating that the development and launch of a scientific satellite was not "to interfere in any way with the ballistic missile program."[28] In effect, this meant that it could not be launched on a vehicle intended for military purposes. When American scientist James Van Allen responded to the Soviet challenge to launch a satellite as part of the IGY, the question for the U.S. government became which launcher to publicly develop and devote to launching a scientific satellite.

Wernher von Braun, the foremost launch expert in the United States and the developer of the Redstone launcher, was working for the Army in Alabama. Interservice rivalry for the generous funding that space was bringing to the military had already created the need to divide the mission between the Army and the Air Force, so that each could get a share of the funding pie. The Air Force got the strategic mission, meaning the long-range, intercontinental launchers, while the Army got the tactical mission, meaning the intermediate-range launchers. Von Braun was disappointed and livid. Throughout his career, he had longed to work on spaceships, not weapons. Now, he was designing not only weapons, but tactical weapons. His launcher design, the Redstone, was clearly the most advanced of those in development, but it was being developed for use as an IRBM.

The only other launcher under development was a Navy Viking/Vanguard launcher, specifically being built to launch a scientific satellite. Based on NSC 5520, it was clear that Vanguard should be devoted to the IGY effort. The decision makers knew that von Braun was closer to success, but speed was not the key. In fact, for U.S. satellites to successfully accomplish their reconnaissance missions, they had to be allowed overflight of Soviet territory; otherwise, they could meet the same fate as had the downed U-2 aircraft in 1960. Government lawyers felt that the best way to succeed was to allow the Soviets to go first, and thereby have the Soviets themselves set the legal precedent for satellite overflight. Then, they would have difficulty complaining about it later. So, though the Vanguard launcher was the least developed at the time and was being built by the Martin Company—which was also developing the Titan ICBM, Martin's clear priority—Vanguard was given the nod for the IGY satellite launch in 1955. Von Braun's eagerness to launch first was well know, to the point that guards were sent to Cape Canaveral with von Braun when he tested the Redstone/Jupiter, to make sure that he did not place a payload on board that might "accidentally" wind up in orbit.[29]

According to pioneer cosmonaut Georgi Grechko, in early 1957, the Soviets received word that an American scientist was presenting a paper on October 5, 1957, at a scientific conference in Spain on the launch of a satellite. The Soviets speculated that the paper was intended as a postlaunch debrief. In fact, the question of whether the Americans intended to launch a satellite on October 5, 1957, was posed to the KGB (Komitet Gosudarstvyennoy Bezopasnosti, or Committee of State Security), which replied that there were no indications that the Americans were or were not planning such a launch. With that ambiguity, the order was given to chief designer Korolev to launch something, anything, before that date. Shelving plans for a much larger satellite, Korolev managed to launch the 83-kilogram *Sputnik* on October 4.[30]

The consequent U.S. public panic, drummed up largely by an uninformed press writing articles along the lines of a distressing piece in *Life* ("Arguing the Case for Being Panicky")[31] or another that stated, "Let us not pretend that Sputnik is anything but a defeat for the United States,"[32] caught the Eisenhower administration unprepared. Clearly, something had to be launched, and fast. It was no surprise to many, including von Braun, that the U.S. response to *Sputnik* in December 1957, known as Project Vanguard, exploded before it even cleared the launch pad and ended up being nicknamed "Stayputnik" and "Kaputnik." Soon thereafter, von Braun was put in charge of producing America's response to *Sputnik*, successfully achieved on January 31, 1958, when a Jupiter-C launcher lifted the *Explorer* satellite into orbit.

With the dawn of the U.S. manned space effort in the 1960s, the American public became used to seeing majestic rockets launch heroic individuals into space. Until the space shuttle arrived in 1982, those rockets were largely derived from designs of defunct weapons systems. The early Mercury space capsules, for example, were lifted into orbit using modified Atlas ICBM launchers. The heavier Gemini spacecraft were lifted on modified Titan ICBM launchers. Historically, in the United States, the Soviet Union, and China, missiles have served as forebears of civilian rockets or launchers, not the other way around. The 1999 congressionally mandated Cox Committee investigation, which alleged Chinese technology theft and illegal transfer of technology by American aerospace companies, repeatedly raised alarms about commercial launch technology being used to improve missile technology. But the success criteria for rocket technology are different from those for missile technology, especially regarding reliability.

Missiles do not have to be highly reliable; their deterrent and psychological impact is the same whether or not they have a 50 percent chance of reaching their intended target or a 99.99 percent chance. Being close counts with both horseshoes and nuclear weapons. In any armed conflict, the adversary must assume that at least some missiles find their targets and react accordingly. The driver behind launch vehicle reliability, however, is strictly financial. When a satellite launch fails, the organization that purchased the launch loses the money that it paid (which can be insured), but, more important, it may lose the services of the satellite for several years while a replacement is built. Quite simply, there is no satellite market for cheap but unreliable launch vehicles. It is therefore rather ironic that the *New York Times* and the Cox Committee speculated endlessly during the committee's hearings and in its report on whether or not China's missile program can benefit from increasing the Long March launch vehicle reliability. Whether Chinese missiles are 50 percent, 92 percent, or 98 percent reliable matters little in American threat calculations.[33]

The commercial launch market was originally a U.S. monopoly. The Soviets' unwillingness to extend themselves beyond the Warsaw Pact countries gave the international market, de facto, to the United States. Subsequently, throughout the 1960s and 1970s, all Western communication satellites were launched by American Delta[34] and Atlas rockets. The first flight of the Delta, on May 13, 1960, ended in failure, but this was followed by a remarkable string of twenty-three consecutive successes. Since that time, the manufacturer has evolved. First, Douglas Aircraft merged with McDonnell Aircraft to form McDonnell Douglas in 1967, and, eventually, McDonnell Douglas was acquired by the Boeing Corporation in 1997. Boeing is also a major defense contractor. Boeing's integrated defense systems (IDS) division is, according to its Web site, a $27 billion business with 78,000 employees. Boeing's missile defense systems division is involved with missile defense programs for all phases of a layered ballistic missile defense system—boost, midcourse, and terminal. Boeing is also involved with manned spaceflight, with activity spanning from Apollo to the International Space Station, for which it is the prime contractor.

The U.S. Atlas launch vehicle evolved alongside the Delta, and it too was based on a ballistic missile. On its first 1958 launch, it put the *Score* experimental communication satellite into orbit, but its early history was checkered, suffering fifteen failures in its first thirty-two flights. Like the Delta, the manufacturer of the Atlas evolved. The original contract was with Convair, which was then absorbed into General Dynamics in 1953. In 1994, General Dynamics Space Systems Division became part of Martin Marietta, which also produced the Titan launch vehicle. All are now part of the Lockheed Martin Corporation.

There are two key points to be gleaned from this discussion of aerospace corporate evolution. First, the firms involved have consolidated considerably.

Whereas in the past there were numerous competitors, only two key launch companies, Boeing and Lockheed Martin, remain—and in 2005, those companies formed a 50–50 partnership under the banner of the United Launch Alliance. Though intended to bring down launch costs, it also effectively eliminates competition, at least domestically. Either separately or as a joint venture, both companies are involved in both the commercial and defense sectors. If one sector is down, they must look for opportunities to make up the business in the other. So, for example, if commercial communication satellite launches are slow, or if there is little potential for new NASA activity on the horizon, the companies must look for big-ticket defense programs to generate corporate earnings.

Today, over 50 percent of commercial communication satellites are launched by the French–European Ariane launcher. The United States lost the market because it refused to allow a European communication satellite to be launched on an American vehicle. Initially, the United States prohibited the export of space launch technology, both to support its own launch providers and to prevent the spread of launch, rocket, and missile capabilities. The intent was to make other countries dependent on the United States and maintain control of sensitive launch technology. The French were particularly uncomfortable with that situation, but were initially unable to garner the requisite support of Germany and Britain to develop an independent launch capability.

In 1967, however, France and Germany decided to build two experimental communication satellites, *Symphonie A* and *Symphonie B*, which NASA agreed to launch. But as the launch dates, set for 1974 and 1975, approached, the State Department demanded that NASA require assurances from Europe that it would use the satellites for only experimental purposes and not carry commercial traffic that would compete with Intelsat commercial business. The U.S. interest in preventing European satellites from competing with Intelsat satellites was purely economic: Intelsat, set up in 1964 with strong U.S. backing and majority ownership, was the international organization intended to assume ownership of communication satellites and take responsibility for managing the global system.

The idea of the United States imposing such restrictions on Europe was the ammunition France needed to convince the other European spacefaring nations that Europe needed its own launch capability.[35] The result was the development of the Ariane launcher. Ariane's first launch came on Christmas Eve, 1979. Ariane 4 soon became the world's most reliable, capable—and used—commercial launch vehicle. In 2003, Ariane 4 was phased out in favor of Ariane 5. Originally plagued with technical problems that resulted in failures and a loss of market share, those problems seem to have been fixed, and a number of new contracts have been signed. Attempts by the United States to tightly control sensitive technology—in the case of Ariane, not out of military concerns, but protecting American economic interests—backfired and created a very successful competitor. Unfortunately, as indicated by its export-control laws, the United States seems unwilling to accept this case as a lesson learned, and is charging ahead with the same zeal to control technology today, likely headed toward the same results.

The Difficulties of Controlling Technology

The U.S. government's approach to controlling sensitive technology traditionally has stemmed from asking one fundamental question: What technology should the United States attempt to control? There are two basic answers: either everything that has potential military value should be controlled, or the United States must accept that it cannot control everything and should build high fences around small areas of especially critical technology over which it retains a monopoly (such as stealth technology). If the second approach is taken, it then becomes critical to stay ahead of everyone else through R&D in uncontrolled areas.

Current U.S. policy assumes the first approach, that everything with potential military value should be controlled. A necessary precursor question, however, has been missing from consideration: What does the United States have the ability to control? Consequently, the U.S. approach has been impossible to implement. Some basic electronic components, considered sensitive and subject to control as part of a commercial satellite sale, are commercially available at any Trak Auto store.[36] How to control the spread of and access to technology—how to design and implement effective safeguard mechanisms and penalties for violations—remains unknown. Should Pentagon monitors watch Trak Auto stores for foreign customers? The U.S. government has said that high-performance supercomputers sold to China are to be used for only commercial, not military purposes. The only way to verify that, however, would be through on-site monitoring, in perpetuity. As any parent of a child with a computer knows, perpetual surveillance is practically impossible—in a home, let alone a country with more than 1 billion people.

It can be argued that all technology is in some manner dual use, and the notion of "dual use" is not limited to technology either. Food aid to another country can be argued as dual use if it is diverted to troops and subsequently used to support potentially hostile forces. It is often not the "know-what," or hardware, that is key; rather, it is often "know-how"—which is even harder to both define and control. If a satellite builder suggests to officials at a foreign launch site that clean clothing should be worn when working with its hardware to avoid contamination, has sensitive technology been shared? Or if a satellite builder suggests that a foreigner buy a college engineering or business textbook to overcome language difficulties, has a defense service been rendered? Know-how potentially includes everything from explaining a technical process to merely agreeing with what another party thinks is the reason for a launch failure. In addition to regulating the transfer of goods, materials, processes, and technology, technical assistance is also regulated by the government. The problem is that nowhere in the government regulations, called the International Traffic in Arms Regulations (ITAR, discussed further in chapter 6), is "technical assistance" clearly defined. While ITAR amendments made in 1993 do discuss a technical assistance agreement, ambiguity abounds:

120.22—Technical assistance agreement.

> An agreement (e.g., contract) for the performance of a defense service(s) or the disclosure of technical data, as opposed to an agreement granting a right or license to manufacture defense articles. Assembly of defense articles is included under this section, provided production rights or manufacturing know-how are not conveyed. Should such rights be transferred, @ 120.21 is applicable. (See part 124 of this subchapter).

Clearly, in addition to selling goods, such as satellites, the ITAR concerns itself with selling services. In practice, however, a service turns out to include virtually any discussion regarding a good.

Several times, the ITAR ostensibly attempts to define a service, but without providing much useful guidance. One example: the phrase "foreign defense article or defense service" includes "any non-United States defense article or defense service of a nature described on the United States Munitions List regardless of whether such article of service is of the United States origin or whether such article or service contains United States origin components."[37] Manufacturers must also provide guarantees that both hardware and knowledge cannot be passed on to third parties as well. How that can possibly be accomplished beyond a reasonable degree is left unstated, however. The United States is using a system with rules that are not enforceable in the real world.

There is also an assumption, not necessarily valid, that transfer is in a particular direction. If an American college student has a summer internship as a lab assistant to a French rocket scientist, current policy assumes that the information is going from the student to the scientist and classifies the student's services as a defense service—nonsensical, but true. The further assumption that useful knowledge with direct military application can be transferred by merely mentioning a key phrase or two, visiting a facility, or even working at a facility for a short time is also problematic, but nevertheless perpetuated.

That space technology is largely dual-use technology creates a plethora of difficulties that permeate political and economic issues. The United States does not have exclusive control over most technology and is controlling less and less of it all the time, partially because of its zealous efforts at control. Other countries go to great lengths to avoid having to deal with simultaneously draconian and ambiguous rules for buying U.S. technology, and so are developing alternative industrial sources. In 2003, the French company Alcatel Space was awarded the contract to build the *Chinasat 9* communication satellite. This satellite was proudly advertised as "ITAR free," meaning that no components of American origin were used, and so the satellite was not subject to U.S. export laws. American firms were not allowed to bid on the contract. In trying to control the proliferation of certain technologies, the United States has created circumstances that may allow not only for proliferation of sensitive materials, but proliferation with fewer restrictions than it might have had coming from the United States, and with profits going elsewhere. Such a miscalculation creates a dangerous situation for the United States in the future. The very nature of dual-use technology complicates nearly every discussion of space hardware and capabilities, and countries using or developing it. It also complicates efforts between countries to cooperate on space projects, including those involving manned spaceflight.

Whether recognized and appreciated or not, space and space-related technology is a part of everyday life for people in the United States, and increasingly for others around the world as well. Because space technology is dual use, however, other countries' craving for the benefits to be yielded from space creates concern in the United States. Most recently, those concerns have been dealt with by attempting to deny technology to others, while developing more and more technology ourselves, especially in the military realm. Attempts at such denial in a globalized world, however, are increasingly fruitless. And in attempting to hoard technology while simultaneously expanding its own military space capability, the United States only increases others' desire and determination to acquire the technology that the United States clearly values. Technology denial and hoarding is also seen as part of a pattern by the United States to hold back the development of other countries.

Past attempts to deny technology to other countries have more often than not proved futile, if not backfired completely, as in the case of the launch technology and development of the Ariane rockets. While there is potential for success in controlling technology proliferation, such as by selling older technology or creating reliance on the United States for parts, that option is often viewed as a form of acquiescence, carrying too high a risk of exploitation by those advocating that the United States gains in the security dilemma only by acting on its own and remaining not just superior, but dominant. The increased global commercial availability of dual-use technology products, however, demands further consideration. Otherwise, the United States could quickly end up with 100 percent control of nothing—a security risk that is both unnecessary and counterproductive.

While NASA has been the civilian and public face of the U.S. space program, it has certainly "shared" technology with the military. Until the space shuttle, NASA's launchers were all birthed as military missiles. The cargo bay of the shuttle was specifically sized to allow NASA to carry certain DOD payloads. Many of the component parts and technology associated with NASA's current focus on returning to the moon, and perhaps going on to Mars, have potential value to the military. Because the United States knows from experience that, at a basic level, capabilities can be transferred between the civilian and military space realms, it worries about space programs of any variety in other countries, and wants to protect U.S. military space superiority and establish U.S. space dominance. What it appears to worry less about, however, is the fate of the American manned program.

From Apollo to Where?

THREE

No bucks, no Buck Rogers.
—Fred Ward, as Gus Grissom in *The Right Stuff*

Name three of the Apollo astronauts. Now, name three current astronauts. Naming the Apollo astronauts was likely easier. The Apollo program, which culminated with the United States successfully sending twelve men to the moon and safely returning them to earth between 1969 and 1972, represents a glorious part of American history. Neil Armstrong was the first person to step off planet earth and onto another celestial body. That event was not only a shining moment for Americans, but a spiritual moment for all mankind.

Nonetheless, by the time Armstrong walked on the moon, funding for the Apollo program was already being cut. The United States had accomplished its goal, the foreign policy objectives behind the program had changed, and commitment to manned space began to falter. NASA's image and, subsequently, the U.S. image of commitment to space exploration and space for all mankind had served both NASA and the United States well, and was thus perpetuated, but it began to be more symbolic than reality. America's commitment to military space, however, has been consistent, and, with the exception of a period during the Clinton administration when questions about program management and direction arose, funding for military space efforts has grown steadily. With Apollo as its shining moment, it is now unclear what the future of the manned program will be, to the detriment of the United States.

If where money is spent represents true organizational or national priorities, then it is clear that the United States is more interested in military space projects than in civilian ones. Declared military space spending has spiked since 2000, from approximately $13 billion annually to almost $25 billion annually, and that likely represents a conservative estimate. According to Marcia S. Smith, "The Department of Defense (DOD) has a less visible but equally substantial space program. Tracking the DOD space budget is extremely difficult since space is not identified as a separate line item in the budget. DOD sometimes releases only partial information (omitting funding for classified programs) or will suddenly release without explanation new figures for prior years that are quite different from what was previously released."[1] During the same period, NASA's budget has increased to about $16 billion. Space programs are inherently expensive, and with a limited amount of money to be spent, NASA has gone from being the fair-haired child to the stepchild in priority.

Perhaps because of its relative youth, the United States is a crisis-response society. It takes a lot to rouse the United States, but as Osama bin Laden and Saddam Hussein have learned, once its wrath is unleashed, it is a fierce and formidable opponent. Unfortunately, Americans also seem to lose interest quickly.[2] Politically motivated to go to the moon by the Soviet launch of *Sputnik*, in less than a decade, the United States was successful beyond anyone's wildest imagination. Once there, however, we basically looked around and returned home, and since then, except for a few die-hard space enthusiasts, the American public shows more interest in its space museums than in space exploration. Meanwhile, the government is focused on using space for military purposes. The United States once held attention and admiration from abroad for its manned space achievements, and the rest of the world eagerly sought to join the adventure. Now, we are in danger of ceding the leadership position that the United States has held in manned space programs since Apollo—with clear negative implications for broader security relations.

The National Air and Space Museum in Washington, D.C., holds the Guinness Book of World Records title for a museum hosting the most visitors in one day: 118,437 on April 14, 1984. Americans take great pride in their space achievements. But opinion polls have consistently shown that the public regards space exploration as expendable compared with other demands for federal funding, such as health care, education, schools, and defense. This was true even at the height of the Apollo era, and since then, the manned space program has been struggling to find a raison d'être salable to both politicians and the public.

Space was a nonissue in the 2004 presidential campaign, and considering ongoing problems in Iraq, a skittish economy, concerns with homeland security, and other issues facing the country, that is not surprising. Even when asked, however, the Bush campaign failed to respond to inquiries about why the president had not talked about his vision, announced in January 2004, for the United States to return to the moon and travel to Mars and beyond.[3] After the last presidential debate, there was a "space debate" in Washington, in which supporters spoke out on their candidates' space records, but few outside the space community knew or cared about the debate. The Aerospace Industries Association (AIA) tried to raise candidates' interest in their concerns by rating the candidates on seven key industry issues, including overhauling the air-traffic-control system, loosening restrictions on exporting high-tech products, increasing space-based surveillance and imaging assets, committing to replacing the aging shuttle, and funding research. Bush was rated "somewhat higher" than Kerry overall, simply because he had spoken on more of the issues than had Kerry. Given the deafening silence from the Kerry camp, however, the Bush response was like being the tallest building in Topeka. The report went on to say that Bush administration officials "made a lot of promises but haven't delivered,"[4] particularly with regard to rationalizing the export control process. Follow through, however, will be absolutely essential if NASA and the U.S. manned space program are to have a future, let alone maintain their leadership positions in the world.

Meanwhile, NASA is caught between a rock and a hard place. Created in 1958 and responsible for America's single most glorious technological achievement within eleven years of its inception, NASA is nevertheless a government bureaucracy, and bureaucracies are not usually risk takers or creative, because they exist at the pleasure of others. It is in the organization's best interest to not rock the boat or even draw too much attention to itself. Additionally, most bureaucracies are responsible for managing the work of others, rather than actually doing the work themselves. The Department of Labor does not build buildings, pave roads, or clean hotel rooms; the Department of Education does not teach classes to students in public schools. However, NASA is largely responsible for developing and operating technology that performs dazzling feats in the harsh environment of space, demonstrating America's global technological leadership on a usually tight budget. In that respect, it is different from many other bureaucracies and is more like the DOD, which both creates and executes war plans.

The scientific data and technology that NASA's work yields have value far beyond exploration. Traditionally, these benefits have been considered as spin-offs, including everything from Tang and Teflon to the development of materials used in airplanes and heart valves. More recently, benefits are not just products, but include the use of space hardware in everyday life through, for example, the use of GPS and communication satellites.

However, as mentioned earlier, Americans' support for space exploration itself—and manned space exploration in particular—is complex. Government and privately commissioned studies nearly unanimously support manned space exploration,[5] and public-opinion polls show strong or at least moderate support for it.[6] However, when poll respondents are asked to rank "space" among other federal spending priorities, it becomes clear that the support is more rhetorical than fiscal. Would that change if the United States faced losing its leadership in manned spaceflight? Many local newspapers seem to feel that there is already a new manned space race to the moon and that the United States is losing to foreign competitors, such as Europe, Japan, India, and China.[7] These reports often exaggerate the capabilities of other countries or underestimate those of the United States, but other countries clearly have recognized the value of space, including manned programs, and are not waiting for the United States to create a viable plan for future activities and invite them to join. Let there be no mistake: the negative strategic effects of the United States ceding its manned space leadership would ripple through American political, economic, and technological communities.

The United States must not allow its leadership role in manned spaceflight to slip away. Manned spaceflight requires pushing the envelope in areas of science and engineering otherwise potentially neglected, such as in the fields of space medicine and life-support-systems engineering. The direct benefits of a particular program to the economy or to defense may not always be identifiable in advance. The GPS was once a government program without a clear mission, but it has certainly demonstrated that we should not be bound by the limits of our imagination. Study after study has been unable to provide an economic rationale for manned space exploration in advance. That should not be surprising. When Lewis and Clark explored the American West, the official reasons for their journey were to show the American flag and to locate beaver dams, as beaver-skin hats were the current fashion rage. Few would argue, however, that beaver-skin hats were the primary economic benefit of opening the American West. The lesson of the Lewis and Clark story is that you often do not know what economic benefits are to be reaped from exploration until after the fact. The importance of space in building scientific capability generally should not be ignored, either. It certainly is not elsewhere. China is acutely aware that it has a long way to go toward becoming a science "power," and it hopes that human spaceflight will accelerate its movement up the learning curve. For the United States to maintain its strength in science and engineering fields, it is therefore imperative that it stay active in space. Finally, while the world understands that the United States places a high priority on military space capabilities, manned space activity represents the future. Space programs represent imagination and vision, and as the world's current leader in space, the United States must remain so into the future. Apollo took us there originally, and the model still offers considerable lessons to learn from.

Apollo and Its Aftermath

Popular history tells us that the Apollo program exemplified NASA's "can-do" attitude and the visionary perspective of the Kennedy administration. If only, some space-exploration advocates still wistfully muse, another U.S. president possessed such imagination and vision, the glory days of NASA space exploration would return. Those reflections are about half-right. The popular historical view of Apollo rightly glorifies NASA's spirit, but greatly embellishes the clarity of its vision.

"*Sputnik* shock" was far more traumatic for Americans than the Eisenhower administration had anticipated. Americans were quickly convinced, largely by uninformed media analysts, that if the Soviets could put a satellite into orbit, then they could do the same with nuclear weapons. More generally, the perception spread that the United States was technically inferior to, and hence potentially weaker than, the Soviet Union—an early example of a techno-nationalist appraisal. That perception took on a global dimension. *Sputnik* created a crisis of confidence because, in the Cold War, nuanced changes in the Eastern and Western blocs were important, even if they were often based on nothing more than perceptions of which bloc could develop the most sophisticated technology. Consequently, NASA was created in October 1958 as a civilian, peaceful space organization, to contrast to the militaristic Soviet program.

However, the United States already had an active military space program under way in 1958. But the Kennedy administration recognized the symbolic power and allure of space accomplishments. It realized that the United States could use space to garner prestige and counter the negative perceptions generated by *Sputnik*'s success. Further, the technology needed for civilian space activities would also clearly benefit military activities and the U.S. economy. If the technical risks could be managed, the benefits were potentially enormous. On one level, space became a Cold War battlefield, in which scientists and engineers were the soldiers at the front lines, fighting for the prestige and global influence that would flow from technical prowess. On another level, the knowledge and hardware created would bring domestic benefits to the symbolic and military arenas and beyond.

Subsequently, within three years of NASA's inception and a mere twenty days after Alan Shepherd became the first American in space on May 5, 1961, Kennedy announced the creation of the Apollo program on May 25, 1961. As a program intended to take a man to the moon and return him safely within a decade, it was a dramatic gesture to recoup the prestige and consequent leadership status lost after *Sputnik*. First and foremost, Apollo was a strategic leadership program. Internationally, prestige was expected to translate into Cold War influence, and it did. But there were multiple other returns on the government's investment in Apollo as well. Education and on-the-job experience for Apollo scientists and engineers created a generation of highly trained technical personnel. Engineering programs were specifically set up in colleges and universities to meet the need for new and specialized aerospace skills. Economically, the benefits to the United States of the space race generally, and the Apollo program specifically, were far reaching, both direct and indirect. Tom Hanks's character in the movie *Apollo 13* shows a congressional delegation through the Vehicle Assembly Building at the Kennedy Space Center. Those tours were once a regular NASA function. Escorts pointed out which parts of the program were produced in which state of the union, and how the dollars spent on Apollo were being spread across the country. This information was politically necessary to keep the funds flowing. Government money was expected not only to get a man to the moon, but to employ a great many people in the process. As crass as it may seem, jobs, especially in key electoral states, such as Florida, California, and Texas, have consistently been key factors in politicians' decisions to support space programs, from Apollo to the space shuttle and the International Space Station.

At its core, however, Apollo was a government program, and programs are activities funded to support policies. The policy behind Apollo was not exploration, but part of a leadership strategy in the Cold War fight against the Soviets, not for visionary exploration goals, or the development of Tang, Teflon, Velcro, or even better missiles. Apollo harnessed soft power for the United States. At the end of the 1960s, when the competition ended with détente and the United States wanted to reach out to the Soviets rather than "beat" them, the policy motivation disappeared, and so did program support and funding. The last three Apollo flights (*Apollo 18*, *19*, and *20*) were canceled.

The difference between Apollo and all subsequent manned programs is that none since has had a strategic goal. Subsequently—from the shuttle, to dreams of a permanently manned space station, to President George H. W. Bush's Space Exploration Initiative—they have all stumbled, lived life on the edge, suffered a death of a thousand cuts, or failed outright. The lesson of Apollo is simple: without a strategic purpose, manned spaceflight is not deemed sufficiently important to warrant the kind of government resource investment necessary for success.

Science is interesting to scientists and appreciated by others who see and recognize its benefits. But that is not always the case for politicians and some of the public, as illustrated by the congressman mentioned in the previous chapter who was not interested in funding a weather satellite when he could already watch the Weather Channel. As the saying goes, nobody throws a parade for robots. Manned space activity captures the imagination in ways that machinery rarely does, but it is also on average ten times more expensive than is unmanned spaceflight, due to using systems with much higher safety margins to "man-rate" the spacecraft, and the including complex life-support systems. The argument is absolutely valid that more science can be done using unmanned systems, but there is rarely the requisite public interest and, subsequently, political support, for financing expensive unmanned science programs. There are exceptions. The NASA Pathfinder Mission to Mars in 1995 held Americans spellbound with its near real-time images from the planet's surface. Images from the Hubble Space Telescope have also captured the public's imagination. For the most part, though, people like to watch people. And, generally speaking, when funding rises for manned space, it also rises for unmanned programs.

For NASA, its bureaucratic birth to fulfill a strategic mission created a legacy with which it still struggles. Scientists are not always the most politically astute people, and those at NASA, at the outset, believed that exploration was their mission, even though the organization had been created in response to *Sputnik*. Uncharacteristically for a bureaucracy, NASA was originally budgeted for more money than its plans required, as its scientists and engineers developed an incremental space program, focusing first on unmanned scientific missions and intending to move to manned spaceflight if and when needed. When T. Keith Glennan, NASA's first administrator, told Eisenhower of NASA's careful plans for scientific exploration, the president pointed out of the Oval Office window toward the moon and said, "You know Keith, its been there for a long time."[8] The implication was that science exploration could wait and that NASA needed to focus on restoring America's technical credibility and world image. In time, NASA management understood, and began the work that culminated in the Apollo moon landings. NASA employees, mostly scientists and engineers, were thrilled that they were being tasked to do things that they all loved to do anyway—respond to a great scientific and technical challenge—though for the most part they were oblivious to the policy behind their tasking. But as mentioned, when the policy of beating the Russians changed, and the reason for the programs vanished, so too did NASA's reason's for being. Most NASA employees outside of management did not know or understand this.

When NASA's mission was no longer linked to a strategic goal, it began to struggle. By then, NASA's legendary can-do culture had been established, but it was a culture that knew nothing other than nearly unlimited funds and favorable treatment from the government and the public. It had no experience or aptitude in being treated just like any of the other junkyard-dog bureaucracies scraping for survival money. With the can-do spirit came an organizational pride verging on arrogance, cited as an issue in both the post-*Challenger* and post-*Columbia* investigation reports.[9] Further, and perhaps most important, NASA no longer had a clear or supportable mission. Failure to validate or change NASA's mission was the beginning of the failure of national leadership regarding manned spaceflight. After Apollo, NASA should have either had its mission clearly redefined in accordance with realistic budget expectations, or been given a sufficient budget to continue the official reason for its existence—exploration—even though the real reason, beating the Russians, had disappeared. Neither was done.

Alternative futures for NASA have been suggested over the years, including turning it into an earth defense agency, more active in defending the planet from threats such as asteroids and climate change; becoming a space-business incubator, prioritizing commercialization; and becoming a space science agency, emphasizing robotic exploration. Any one of these might be worthy goals, if the perceptual leadership ramifications of aborting or minimizing manned spaceflight were considered acceptable. That debate has not occurred, however, nor has serious consideration been given to simply folding NASA activities into DOD; having NASA as the public face of U.S. space activities is still considered important enough to warrant its existence. Instead, successive administrations and Congress have allowed NASA to flounder, and NASA's culture has exacerbated the problems that lack of leadership inherently created.

The real culture shock to NASA came with the Post Apollo Program (PAP). Even though the message should have been clear after the last three Apollo missions were canceled due to lack of interest, NASA still made ambitious plans to build space stations and lunar posts, and go to Mars. What President Richard M. Nixon approved, however, was the smallest piece of the smallest option presented to him as potential PAP plans—a space shuttle to serve as a taxi to a space station. The space station itself was put on indefinite hold. This meant that NASA was to build a reusable space vehicle, though it had nowhere to go. While Nixon was personally enamored with space flight, he was first and foremost a pragmatic politician. Approval of the shuttle provided valuable jobs for voters before the midterm 1972 elections. For NASA, there was a catch. For the first time it was tasked to develop the technology and build the very first reusable space vehicle—and show that the vehicle was cost effective.

How does one estimate the cost of building something that has never been built, with materials that may not have been invented yet? The Vanguard rocket program, the first civilian U.S. space project intended to take a U.S. satellite into orbit as part of the IGY, was initiated in 1955 before NASA was even born. Its original cost estimate was $10 million. That quickly jumped to $15 million, then $63 million, and then $99 million. Estimation has improved since then, though. Early in the Apollo program, then-NASA administrator James Webb was asked to testify before Congress about how much the NASA program would cost. He tasked his best accountants and engineers to estimate a figure, but he thought that their initial estimate was too low and sent them back to recalculate. As he was walking out the door of NASA headquarters to go to Capitol Hill, he was handed the new figure. At the last minute, sitting at the table before the congressional committee, he decided to double the amount given him by his experts. He told Congress that the Apollo program would cost approximately $25 billion. Apollo cost almost exactly $25 billion. The rule of thumb in calculating the cost of developing space technology—and this should be remembered later, when estimated costs for developing a new crew exploration vehicle (CEV) or missile defense are cited—is to take the high estimate, and double it. The Congressional Budget Office stated in a September 2004 report that historically, NASA spaceflight programs end up costing about 45 percent more than originally estimated, while Air Force and missile defense programs cost closer to 69 percent over their estimates.[10] Clearly, however, these are averages. The more cutting-edge the program, and the greater the need for integration with other programs, the higher the potential increase.

The change in attitude regarding funding between Apollo and the shuttle points out a problem that perpetually plagues space programs, civilian and military, though it is more often and more easily overlooked in military program cases. Space developers must underestimate the costs of building their cutting-edge, first-time-ever technology, or Congress will have sticker shock and fund nothing at all. This is especially true regarding purely scientific or exploration programs. In 1989, President George H. W. Bush stood on the steps of the Air and Space Museum on the twentieth anniversary of the *Apollo 11* moon landing and announced his SEI program for a manned return to the moon and going on to Mars. NASA came up with a realistic estimate of between $400 billion and $700 billion to implement the program, and it died within a year.

The film *Apollo 13*'s example of astronauts showing members of Congress how Apollo brought jobs and benefits to their states, and the change in attitude exhibited post-Apollo, illustrate the often complicated relationship between Congress and space. Unless facilities such as a NASA field center (of which there are ten)[11] or an aerospace industry are located in their districts, members of Congress do not usually prioritize space programs. While not necessarily adverse to space activity, their funding prioritizations reflect the concerns of their constituents. Funding decisions are made in congressional subcommittees; for many years, the congressional subcommittee that funded NASA was also responsible for funding veterans' affairs and the Department of Housing and Urban Development. Clearly, more voters were most immediately concerned with veterans' benefits and housing over spaceflight, making NASA the subcommittee's stepchild.

In 2005, at the instigation of House Majority Leader Tom DeLay (R-Tex.), Congress reorganized the committees responsible for the NASA budget. In the House, NASA was placed with the Departments of Commerce, Justice, and State for appropriations subcommittee consideration, under the leadership of Frank Wolfe (R-Va.). The Senate quickly followed suit, moving NASA to the Commerce, Justice, and Science Subcommittee of the Appropriations Committee. Initially, NASA supporters saw this as a good move for them, but it may not turn out that way. Like NASA, State and Justice have lots of budgetary holes to fill.

Further, members of Congress have long viewed the NASA budget as a great opportunity to include pork projects in their districts, sometimes at the expense of NASA needs. In 1992, Jamie Whitten (D-Miss.), chairman of the House Appropriations Committee, wanted to continue work on a new NASA booster-rocket factory being built in his district. However, NASA no longer supported the project and cut it from its budget to support other priorities. Whitten used his influence to get $360 million for the unrequested solid rocket motor project back into the NASA budget, cutting from funds allocated for shuttle spare parts, tracking and data-relay satellites, and unmanned rockets to do so.[12] This is not out of the ordinary.

Members of Congress have also used NASA programs to carry out other agendas. More than once, human rights issues have been raised in conjunction with considerations for cooperating with other countries on space programs. The idea is that space cooperation will be held out as the carrot to get other countries to change their domestic political systems in accordance with U.S. views. This has never been an effective approach, either during the Cold War concerning the Soviet Union or afterward. Countries see human rights as a domestic political issue, a matter of sovereignty. Trying to make a direct link between human rights and space is unproductive. Moreover, given the Abu Ghraib prison debacle, and with Amnesty International citing the U.S. prison at Guantanamo Bay, Cuba, as having its own human rights issues, U.S. demands that other countries adhere to human rights standards—most often directed at China— are seen as simply hypocritical.

The point is that Congress is not always helpful in providing NASA with a rational way to proceed in its efforts. Some members are diligent stewards. Others are willing to point out NASA's flaws, but unwilling to either give consistent direction or the financial support needed to fix the flaws. Even during Apollo, Congress began to cut funding long before Neil Armstrong stepped onto the moon. Only commitment from the office of the president—first Kennedy, and then Johnson—maintained because of the program's strategic goal of beating the Soviets, ensured that Congress maintained support for long enough to guarantee success. With the space shuttle, there was no such commitment or strategic purpose.

Therefore, technically, there were only two options for NASA regarding the shuttle: either spend whatever funds were necessary to develop a vehicle with low operating costs, or spend less on development and end up with a vehicle with higher operating costs. Congress made it clear that development money would be very limited and it wanted cost efficiency, or there would be no funding whatsoever—so there really was no option. NASA could not design a cost-effective vehicle. In fact, the shuttle is anything but cost effective, because it jettisons a portion of its hardware, the external tank, into the Atlantic Ocean after each launch. The only way NASA could reconcile low development funding and high operating costs with cost efficiency was to show high vehicle usage, amortizing (and consequently lowering) the operating costs over many flights. So, NASA made flights-per-year projections for shuttle use that, in retrospect, were clearly unachievable. Shuttle flights cost about $500 million each, and the shuttle has been grounded three times, twice after catastrophic failures, and again after return to flight in 2005 when more problems were identified. Space historian Alex Roland has referred to the shuttle as "the world's most expensive, least robust and most deadly launch vehicle."[13] But NASA built what it could afford. The agency paid for the shuttle virtually on its own, though European Space Agency and Canadian partners did contribute auxiliary hardware.

International Cooperation in Space

Space has always been a venue for both cooperation and competition.[14] Apollo was a national program to "beat the Russians." But the Apollo-Soyuz Test Project in 1975 used one of the last Saturn 5 rockets built for travel to the moon to dock with the Soviets in LEO as a show of U.S.–Soviet cooperation and friendship: strategic communication at its finest. Many scientists saw the highly publicized handshake-in-space mission as wasteful, even though the astronauts conducted multiple scientific experiments while in orbit. Politicians got everything they wanted from the mission, specifically a well-publicized demonstration of improved U.S.–Soviet relations.

The Space Act of 1958, which created NASA, states that the agency's objectives include "cooperation by the United States with other nations and groups of nations in work done pursuant to this Act and in the peaceful application of the results thereof."[15] Generally speaking, 1958 to 1969 was a golden age of cooperation, with the United States eagerly reaching out for partnerships. The motivations of the United States were both altruistic and pragmatic. Altruistically, as the leader of the free world, the United States was reaching out to share the destiny of stepping off the planet with other countries. Pragmatically, NASA required sites in other countries to track their spacecraft in orbit. Having such a site became a status symbol for some countries—another example of techno-nationalism—making it a benefit for both participants. Cooperation was also seen to promote international economic development, and create new markets for U.S. communications and aerospace products, though ironically, the United States would later deliberately prohibit many of the same products from being exported. A 1987 NASA task force on international cooperation in space acknowledged that, first and foremost, cooperation was about politics: "International cooperation in space from the outset has been motivated by foreign policy objectives."[16] Cooperation has historically been a useful tool for the United States to shape other countries' space programs in accordance with U.S. interests, as the United States co-opts both other countries' program directions and their limited resources. During this initial period of cooperation, beyond hosting tracking site locations, cooperation for other countries basically meant participating in a U.S. space experiment or, less frequently, having the United States launch their spacecraft.

There are many different types of cooperation, and some are considered easier than others. Initially, having one country launch an experiment or satellite for another country was deemed to allow relatively clean interfaces, and the United States thus encouraged it. In a clean interface, discrete pieces of hardware are built, minimizing the potential for technology transfer. That premise formed the basis for many cooperative programs and developing the commercial launch industry. Since the late 1990s, however, the United States has considered even these types of activities to be potentially risky.

There is also simple coordination of activities, in which each country or organization builds its own hardware, with the intent of increasing the potential yield when using the hardware together. That type of cooperation has also been considered as a way to minimize the potential technology transfer involved with cooperative programs involving dual-use technology.

Joint development of technology, however, has always been more difficult, not only because of technology transfer issues, but due to the potential for problems with technical interfaces (for example, metric measures used in Europe and Japan versus nonmetric used in the United States). There are also difficult issues with joint development regarding funding. No government official wants to send money to another country to help its economy, so questions about where and how to build the equipment inevitably occur.

Experience over the years with multiple programs and multiple types of cooperation have provided valuable "lessons learned" regarding when and how each of the different types work best. The American Institute of Aeronautics and Astronautics (AIAA) has held a series of workshops on space cooperation, looking at not just the challenges of cooperation, but solutions. The first meeting, held in 1992, focused on "Learning from the Past—Planning for the Future." The second, in 1994, then moved to "Getting Serious About How." The third workshop in 1996 considered "From Recommendations to Actions," and the fourth in 1998, dealt with "New Government and Industry Relationships." The fifth workshop was officially designated as one of the preparatory activities for the United Nations–sponsored UNISPACE III Conference, held in Vienna in July 1999, and looked at "Solving Global Problems." The sixth workshop, "Addressing Challenges of the New Millennium," was held in March 2001, and the seventh, "From Challenges to Solutions," was held in 2004. Through these workshops, models, levels, criteria, and rules for what type of cooperation best supports different goals with different partners have been developed.[17] A first step in any model is to ensure that all partners have a vested interest in success, that all partners fully understand their roles, and that the science and engineering goals are meaningful.

These workshops are sponsored by several national and international organizations, including the UN, and attended by space professionals. The reports generated from these workshops look for ways for nations to cooperate on space programs. They are not partisan; they neither bash the United States nor view it as faultless, and the capabilities and constraints of other countries are reasonably assessed. The latest report, from the seventh AIAA workshop, was released in May 2004. Among its findings was that "successful partnership cannot be built on a weak foundation. Long-term productive cooperation cannot be built on wishful thinking: it requires realistic analysis on the level of commitment that can be sustained by each potential partner." Regarding who is an appropriate partner, the report states: "International cooperation in space should be seen without limitation on number of partners. Space is always seen as a matter of pride for a nation or a region. Cooperation in space activities contributes to peaceful international relations and to generating economic welfare that contributes to some sustainability."[18] Cooperation, however, does not necessarily lower the cost of space programs; the overall cost of a cooperative program is likely to be higher than it would be if the program were the sole responsibility of one country. A rule of thumb is that overall cost increases by about one-third, due to management and interface expenses. Communication channels must be established; technical and legal teams assembled and exchanged, often for prolonged periods (all of the ISS partners have long had offices at Johnson Space Center); and hardware built to specifications compatible with other hardware, and transported. However, cooperative programs should also have greater capabilities, because more partners are contributing and the cost to individual countries to access those capabilities will be proportionally less. The technical rationale for cooperative programs is to make one plus one yield more than two.

Between 1970, with political interest in Apollo waning, and 1984, when the ISS program was first announced, the tone and type of cooperative opportunities began to change. Other countries, specifically European countries, Japan, and Canada, had developed their own space capabilities and so could participate on a more sophisticated level. But with maturing capabilities came increased potential for competition as well. The International Solar Polar Mission (ISPM) was planned during this period.[19] That program was intended to send two spacecraft, one from the United States and one from the ESA, to the solar poles to perform unique (stereo) photography possible because two spacecraft were involved. As it turned out, the mission became known as much for the political lessons that other countries learned regarding the risks and parameters of working with the United States as it did for scientific data yielded.

Quite simply, some years into the program, the United States canceled its spacecraft due to NASA budget cuts. The ESA, already building its spacecraft and now unable to perform the planned mission without NASA's promised contribution, was shocked and felt betrayed. After months of subsequent negotiations, it became clear that there was nothing Europe could do to force the United States to follow through on its commitment. The annual budget process in the United States means that Congress has the ability to allocate money for programs regardless of international commitments. While NASA could have reallocated money from other programs, there was also no obligation on the part of NASA to fund international programs over purely national programs, and in this case, NASA chose not to support the project. The European spacecraft, subsequently renamed *Ulysses*, went off to the sun on its own, with a changed scientific mission. The harsh lesson learned by the Europeans, and carefully noted by other countries, is now implicitly understood in negotiations on other high-budget, long-term cooperative space and other programs with the United States. The key lesson learned was that no legal mechanism exists to obligate the United States to spend money. About the best that can be hoped for to get the United States to abide by its commitments are high-level U.S. announcements that can be used to embarrass the United States, if it later tries to pull out.

The parameters of partnerships also became more clearly defined between 1970 and 1984. While space capabilities and industries were maturing in other countries, they were certainly not equal to those of the United States, nor were the resources for space programs available in or from other countries. What, then, did partnership really mean? From the U.S. perspective, it was contributing the majority share of the money on cooperative programs, and had both the largest and the most experienced aerospace industry. Therefore, while the United States wanted other countries to participate, it required both decision-making control and U.S. manufacture of all critical parts, or initial operating capabilities (IOC). Congress was not going to spend American tax money abroad and rightly so, because there was higher confidence in technology built in the United States than in that built elsewhere. Other countries participated in these space programs to climb the technology development curve and develop their own aerospace industries, for jobs and ultimately economic development. During the shuttle program development, it was not uncommon to see complaints and resentment from Europeans building the pressurized space lab module for optional use with the shuttle about the United States perceiving and treating them as incompetent subcontractors. For the ISS, complaints have been focused more on being excluded from the decision-making process, especially in redesigns. Acknowledging that the United States is the majority partner, the other partners nevertheless have felt that consultation is a requisite and reasonable expectation in any partnership. The many lessons learned from past cooperation validate the value of cooperation.

Running on Inertia

After building the shuttle on a shoestring and living with the resultant technical problems, President Ronald Reagan announced in 1984 that a space station would be built, initially called Alpha, then Freedom, and later renamed the International Space Station—finally providing the shuttle with a destination. The first module was launched in 1998, fourteen years after Reagan's initial announcement. Until then, the shuttle had been a high-cost, relatively difficult to use trucking service to orbit, yet it had formed the centerpiece of NASA's existence. To say that commitment to the ISS was lacking between its announcement and the first module being launched is an understatement. In 1991, the House Appropriations Committee first recommended program cancellation. That recommendation came from the committee chair, Robert Traxler (D-Mich.). Representatives Jim Chapman (D-Tex.) and Bill Lowry (R-Calif.) offered an amendment to the NASA appropriations bill to continue the program, with support for civilian space programs again being driven more by geography than anything else. The amendment was adopted 240–173, and the program survived. That was the first of twenty-two specific congressional votes between 1991 and 2001 (fourteen in the House, eight in the Senate) on NASA funding bills regarding whether to terminate the program. Most objections to the space station stem from the cost overruns that have plagued the program from the beginning. Originally estimated to cost $8 billion, by 2004, its price tag exceeds $25 billion, though nobody is quite sure by how much, and that money funded a significantly scaled-back version of what was originally intended. Each time the station was threatened, however, Congress voted to continue it.[20] On one of those occasions, it missed cancellation by only two votes. During several of these near-death experiences, the only reasons that could be mustered for continuing work on the station were domestic jobs and not wanting to renege on such a high-visibility international commitment.

By the time the first ISS module was launched in 1998, there had been at least seven major technical redesigns. Initially, the ISS was a partnership among the United States, eleven European countries working together through the ESA, Canada, and Japan. Some of the redesigns reduced the size and capabilities of the ISS, much to the chagrin of the partners who sometimes were not included in the redesign decisions, but were affected by them. Other redesigns were to include new partners, specifically the Russians, who joined the ISS program in 1993 at the invitation of the United States, also without consultation with the other partners.

The official reasons for building the space station have changed along with the station blueprints and according to whom one asks. For NASA, it has always been first and foremost about continuing manned exploration and establishing a permanently manned presence in space. Even after repeated cuts in ISS power supplies and equipment, the science that can be accomplished on board is significant. But few Americans are interested in funding a $25 billion science project. In terms of a permanently manned space station, the Russians jumped ahead of the United States in that area with their Mir space station, and NASA was left playing catch-up. In a 1998 letter from Barbara Larkin, assistant secretary for legislative affairs in the Department of State, to James Sensenbrenner (R-Wis.), chairman of the House Science Committee, regarding questions about the ISS, she stated State's position: "The U.S. is seeking to keep Russia constructively engaged in the international arena and perhaps more importantly, Russian participation in the ISS plays a vital role in our non-proliferation program."[21] The ISS has been at least in part a nonproliferation program, keeping otherwise out-of-work Russians scientists employed, and therefore out of potential (missile) mischief. Again, programs support policies. Although programs can theoretically (maybe even ideally) support several policies, when there are questions about which policy a program is supporting, or the validity of the policy, funding support begins to crumble. It becomes a program of interest only to those who directly benefit from the jobs generated. That has been the history of the ISS. If it had not been for domestic jobs and international politics, there would be no ISS.

The Bush Space Vision

On January 14, 2004, President George W. Bush outlined his plan for a new space program, starting with a return to the moon, and with an eye toward reaching Mars and beyond:

> Today I announce a new plan to explore space and extend a human presence across our solar system. We will begin the effort quickly, using existing programs

and personnel. We'll make steady progress—one mission, one voyage, one landing at a time.

Our first goal is to complete the International Space Station by 2010. We will finish what we have started, we will meet our obligations to our 15 international partners on this project. We will focus our future research aboard the station on the long-term effects of space travel on human biology. The environment of space is hostile to human beings. Radiation and weightlessness pose dangers to human health, and we have much to learn about their long-term effects before human crews can venture through the vast voids of space for months at a time. Research on board the station and here on Earth will help us better understand and overcome the obstacles that limit exploration. Through these efforts we will develop the skills and techniques necessary to sustain further space exploration.

To meet this goal, we will return the Space Shuttle to flight as soon as possible, consistent with safety concerns and the recommendations of the Columbia Accident Investigation Board. The Shuttle's chief purpose over the next several years will be to help finish assembly of the International Space Station. In 2010, the Space Shuttle—after nearly 30 years of duty—will be retired from service.

Our second goal is to develop and test a new spacecraft, the Crew Exploration Vehicle, by 2008, and to conduct the first manned mission no later than 2014. The Crew Exploration Vehicle will be capable of ferrying astronauts and scientists to the Space Station after the Shuttle is retired. But the main purpose of this spacecraft will be to carry astronauts beyond our orbit to other worlds. This will be the first spacecraft of its kind since the Apollo Command Module.

Our third goal is to return to the moon by 2020, as the launching point for missions beyond. Beginning no later than 2008, we will send a series of robotic missions to the lunar surface to research and prepare for future human exploration. Using the Crew Exploration Vehicle, we will undertake extended human missions to the moon as early as 2015, with the goal of living and working there for increasingly extended periods. Eugene Cernan, who is with us today—the last man to set foot on the lunar surface—said this as he left: "We leave as we came, and God willing as we shall return, with peace and hope for all mankind." America will make those words come true.

Returning to the moon is an important step for our space program. Establishing an extended human presence on the moon could vastly reduce the costs of further space exploration, making possible ever more ambitious missions. Lifting heavy spacecraft and fuel out of the Earth's gravity is expensive. Spacecraft assembled and provisioned on the moon could escape its far lower gravity using far less energy, and thus, far less cost. Also, the moon is home to abundant resources. Its soil contains raw materials that might be harvested and processed into rocket fuel or breathable air. We can use our time on the moon to develop and test new approaches and technologies and systems that will allow us to function in other, more challenging environments. The moon is a logical step toward further progress and achievement.

With the experience and knowledge gained on the moon, we will then be ready to take the next steps of space exploration: human missions to Mars and to worlds beyond. Robotic missions will serve as trailblazers—the advanced guard to the unknown. Probes, landers and other vehicles of this kind continue to prove their worth, sending spectacular images and vast amounts of data back to Earth. Yet the human thirst for knowledge ultimately cannot be satisfied by even the most vivid pictures, or the most detailed measurements. We need to see and examine and touch for ourselves. And only human beings are capable of adapting to the inevitable uncertainties posed by space travel.

As our knowledge improves, we'll develop new power generation propulsion, life support, and other systems that can support more distant travels. We do not know where this journey will end, yet we know this: human beings are headed into the cosmos.

And along this journey we'll make many technological breakthroughs. We don't know yet what those breakthroughs will be, but we can be certain they'll come, and that our efforts will be repaid many times over. We may discover resources on the moon or Mars that will boggle the imagination, that will test our limits to dream. And the fascination generated by further exploration will inspire our young people to study math, and science, and engineering and create a new generation of innovators and pioneers.

This will be a great and unifying mission for NASA, and we know that you'll achieve it. I have directed Administrator O'Keefe to review all of NASA's current space flight and exploration activities and direct them toward the goals I have outlined. I will also form a commission of private and public sector experts to advise on implementing the vision that I've outlined today. This commission will report to me within four months of its first meeting. I'm today naming former Secretary of the Air Force, Pete Aldridge, to be the Chair of the Commission. Thank you for being here today, Pete. He has tremendous experience in the Department of Defense and the aerospace industry. He is going to begin this important work right away.

We'll invite other nations to share the challenges and opportunities of this new era of discovery. The vision I outline today is a journey, not a race, and I call on other nations to join us on this journey, in a spirit of cooperation and friendship.

Achieving these goals requires a long-term commitment. NASA's current five-year budget is $86 billion. Most of the funding we need for the new endeavors

> will come from reallocating $11 billion within that budget. We need some new resources, however. I will call upon Congress to increase NASA's budget by roughly a billion dollars, spread out over the next five years. This increase, along with refocusing of our space agency, is a solid beginning to meet the challenges and the goals we set today. It's only a beginning. Future funding decisions will be guided by the progress we make in achieving our goals. [22]

To reiterate, the speech laid out three basic goals: to complete the ISS by 2010; to return the shuttle to flight as soon as possible to finish assembling the ISS; and to develop and test a CEV by 2008, conduct its first manned mission by 2014, and return to the moon by 2020, then extend the human mission to Mars and worlds beyond.

Does the public support the New Vision for Space Exploration, also known simply as the Bush Space Vision? Does the public even know or care to know about it? What about politicians? When a government program fails to build enough momentum in Washington to warrant a catchy name, and an all-important acronym, the program is in trouble. For any serious program, the acronym is decided first and the name shaped around it before it even hits the streets. Witness the Uniting and Strengthening America by Providing Appropriate Tools Required to Intercept and Obstruct Terrorism, better known as the USA Patriot Act of 2001. It is also not a good sign when the person in charge of a troubled organization, which is in turn charged with implementing a space vision characterized by some as dead on arrival or sketchy at best,[23] suddenly announces that he is leaving, as NASA administrator Sean O'Keefe did in February 2005, just over a year after the vision announcement. It speaks volumes when the statement's originator, President Bush, declines to speak further about the vision—not even mentioning it in subsequent State of the Union addresses.

Flesh began being added to the bones of the murky vision only slowly. No president wants to pull the plug on the U.S. manned space program. But without a salable plan and serious commitment, evidenced by accompanying funding and political cover—neither of which have occurred to date—that may well be a legacy of the Bush administration, intended or not. Although Bush hopes that future administrations will pay the bills for his vision, hope is not a strategy.

How much will the vision cost, and what is the current commitment? Early media speculation estimated the cost at $1 trillion. For its part, NASA estimates have ranged from $97 billion to around $170 billion, depending on what is included and the time period. The nonpartisan Congressional Budget

Office completed a study putting the cost at $127 billion.[24] Where will the money come from? The Bush administration initially requested $1 billion for the program, probably not enough to pay for the PowerPoint presentations needed to sell programs in Washington. Another $11 billion initially is to come from reallocated NASA money, which can be done, but not without resistance. That leaves somewhere between $83 billion and $158 billion to future funding; using the rule of thumb for new technology development costs discussed earlier, doubling the high estimate, the amount would be closer to $300 billion.

With the near-draconian constraints on the budget due to the war in Iraq, Congress initially took out the $1 billion in funds requested for Bush's space vision from the 2005 Veterans' Affairs and Housing and Urban Development appropriations bill that funds NASA. The White House, however, responded by threatening to veto that bill if the NASA budget did not get additional funds to support the vision. In the budget bill that was finally passed in November 2004, NASA got back almost all the $1 billion requested to get the president's vision started. Further, NASA was given "unrestrained" authority to reprogram money within its accounts: It can rob Peter to pay Paul.

In the past, NASA has shown some agility in reprogramming funds. When the shuttle was over budget and behind schedule in the late 1970s, money from science programs was reallocated to the shuttle in what became known in space circles as the "slaughter of the innocents." The U.S. spacecraft in the ISPM was among those killed. But the congressional pushback from that kind of reprogramming can be substantial. The president wants to reorient all NASA efforts toward his vision. Such a reorientation means killing a number of programs already established and yielding benefits to somebody's constituents, returning valid scientific data, or affecting international commitments.

That Congress acceded to the president's wishes and added to NASA funds does not necessarily indicate a dawning of awareness regarding merit. The additional funds were championed by heavyweight Tom DeLay. Many of DeLay's constituents are employees of Johnson Space Center.[25] As in the past, constituent benefit was likely the motivating factor. While DeLay's commitment to supporting his constituents and the president was useful in this instance, similar political commitment—read protectionism—toward other constituents working on other now-threatened NASA programs can be expected as well.

Retired admiral Craig Steidle took over as the NASA associate administrator for the Office of Exploration Systems in 2004 to implement the Bush vision. A former fighter pilot known for executing large and complex programs, he was seen as O'Keefe's pick, as he understood both the technology and the politics of NASA. Almost immediately, his office issued more than 120 contracts to develop items from solar arrays to food crops genetically modified to grow in space. Steidle intended to build a base of support through contracts having ties to one or more NASA centers, creating jobs that would translate into constituent and political support. Before the plan reached fruition, however, Steidle resigned in June 2005, as part of a large-scale NASA personnel shake-up after Sean O'Keefe's replacement came on board.

Michael Griffin succeeded Sean O'Keefe in April 2005. Griffin's impressive biography includes positions at Johns Hopkins University's Applied Physics Lab, and he was an associate administrator for exploration at NASA. He also served as the deputy for technology at the Strategic Defense Initiative Organization (SDIO), the organization in charge of executing President Reagan's "Star Wars" vision. He is an astute program manager with credibility within the national and international space communities, is a veteran of past cooperative programs, and acutely understands Washington politics. Wanting to accelerate technology development schedules, Griffin quickly reversed key elements of Steidle's implementation plan. Griffin's approach draws from that used in SDIO—a "just do it" approach. Griffin's technology acceleration plan was dubbed, by him, as "Apollo on steroids."[26]

The initial Bush vision of space exploration called for astronauts on the moon by 2020, and then going on to Mars. Griffin's plan focuses on human exploration of the moon by 2018, with Mars placed on a back burner. In the new plan, an Apollo-like vehicle carries crews of up to four astronauts to the lunar surface for stays of up to seven days.

Without any further elaboration, that plan inherently requires developing a new launch vehicle. The shuttle is already past its originally intended lifetime and is limited to LEO operations. In response, NASA is creating the CEV, referenced in President Bush's speech. The cost of developing the CEV has been estimated at approximately $5 billion, and another $5 billion for its launcher. Given that development of the shuttle cost between $50 billion and $100 billion in current dollars, depending on what is included in the calculation, $10 billion for a new vehicle to go to the moon seems incredibly optimistic. Initially, too, it was to be funded by a "go-as-you-pay" or "spiral development" method, which in essence says that it will be built when money becomes available. Steidle used the term "spiral development," as it is common to DOD programs. Griffin disavows the term, but the premise of limited funding for development still applies. Further complicating the funding equation, NASA intends to streamline development, production, and operations of the CEV to meet a first crewed test flight as close to 2010 as possible, but no later than 2012—accelerating the dates provided when Bush originally announced his space vision.[27] To some, these directives appear to conflict. With an unlimited budget, the CEV timetable might be possible. But the budget is anything but unlimited. Considering that the new F-22 fighter jet for the Air Force has been in development for over thirteen years, the entire CEV development plan seems unrealistic.

May 2005 was the deadline for contractors to submit proposed CEV designs to NASA. Independent aerospace analysts were underwhelmed by both the originality and the viability of the submissions. The Lockheed Martin design, for example, was described as "the kind of proposal you slap together cheaply at the last minute for a dumb program that you know will be cancelled."[28] Engineers were not necessarily to blame. They were responding to over 6,000 pages published by NASA about the contracts for the CEV, externally summarized as "Apollo with the names of the modules changed to confuse everyone."[29]

The CEV is envisioned as a much needed next-generation spacecraft capable of taking astronauts to and from the moon, and perhaps beyond. The CEV's development, however, was planned assuming that the shuttle would return to flight on time in 2005 and be used to complete the ISS. The money for building the CEV would then come through retiring the shuttle in 2010 and limiting ISS use to only tasks related to returning to the moon, freeing up money previously planned for those programs. With shuttle flights suspended again for a year after its July 2005 return-to-flight following the *Columbia* tragedy, until July 2006, so that NASA could investigate the continuing problem of foam falling from the shuttle's external liquid fuel tank during launch, timetables remain tenuous.

Theoretically, however, while working on the vision, NASA is also to finish the ISS, which is budgeted between $1.5 billion and $2 billion annually. When and how much of that funding can actually be reprogrammed to the vision while still operating the station safely is a concern. Before the president's January 2004 vision announcement, engineers already felt that the ISS was running on a shoestring. Resupply opportunities are limited, and trash storage is a problem on board. Since the *Columbia* tragedy, ISS has relied totally on Russian spacecraft for transport, and as of 2006, the original arrangement whereby Russia provided free rides to the ISS for American astronauts has expired. As a result, NASA now must factor payments for transport to the ISS into its budget before expenses can be diverted to paying for the new space vision.

After President Bush's 2004 speech, the ISS partners, again surprised by a shift in U.S. civilian space priorities, began questioning how much of their own scarce resources to invest in a program that had just gone from being a NASA flagship effort to almost a nuisance. While NASA committed to supporting the ISS for at least another dozen years after Bush's vision announcement, commitment became focused on finishing it and, potentially, abandoning it, with a severely narrowed research agenda along the way. While remaining rhetorically positive, the partners would be remiss not to think in the long term before making further investments. There has already been speculation that perhaps the United States will finish the station and then turn it over to the partners. But would the partners want it? The operating costs of the ISS have always been the unspoken nightmare among the station's participants, with somebody once comparing it to smoker's cough: nobody wants to talk about it, but you know it can kill you. But administration and consequently NASA attention has already turned elsewhere, specifically back to the moon. Meetings have commenced with potential international partners regarding their possible participation in the space vision. While Griffin says that the partners are "gung-ho" about moving on to the moon with the United States,[30] given their past experience regarding "partnerships" with the United States on space programs from the ISPM to the ISS, and the tenuousness of the vision's future, hesitation on their part would be understandable.

Many long-time critics of the ISS support abandoning it. The need to replace the shuttle has also long been recognized. Griffin himself called both the shuttle and the ISS "mistakes,"[31] though he later "clarified"—that is, toned down—that assessment. But is the Bush vision the right effort, being done in the right way, to make happen now what has been impossible since Apollo? Until 2005, it was thought that the shuttle would need to take between twenty-five and thirty flights to finish the ISS, per NASA's commitment to the ISS partners to do so. That number has now been scaled back to sixteen (fifteen to finish ISS, the sixteenth to service the Hubble Telescope) to complete commitments to the partners. Commitments have clearly become fungible. Griffin has implied that he intends to follow Bush's orders to stop flying the orbiters by 2010 whether the ISS is completed or not. That the ISS will not be completed seems as likely as not. Accomplishing twenty-five to thirty flights by the end of the decade would be nearly impossible; even sixteen seems improbable. If shuttle flights stop and no replacement capabilities are on hand, the United States will be forced to abrogate its commitments to the ISS partners and potentially abandon the ISS after spending more than $25 billion to develop it. Recall that President Bush stated in his vision announcement that "our first goal is to complete the International Space Station by 2010. We will finish what we have started, we will meet our obligations to our 15 international partners on this project." If the shuttle flights continue toward completion of the ISS, that potentially ties up CEV development funding, on which all other elements of the vision hinges. How viable and thought-through does all of this sound?

The need to end shuttle expenditures to have even the slightest chance of developing the CEV raises interesting questions. An article in the *New York Times* had an interesting way of portraying the dilemma of phasing out the shuttle while building the CEV. "The question," the article read, "might have come from a caller on a consumer-advice program like 'Car Talk': how much should a driver expect to spend keeping a car running after deciding to buy a much newer, far more expensive and thrilling model that has not rolled out of the factory yet?"[32] Astronauts and Florida lawmakers are already paying attention.

Astronauts have raised questions of safety. Before the shuttle returned to flight after the *Columbia* disaster, they wanted assurance that all the recommendations of the Columbia Accident Investigation Board (CAIB) were fulfilled.[33] The recommendations included ensuring protection of the thermal protection system, as damage to the system was determined to have caused the *Columbia* accident; requiring better imaging capabilities to allow NASA to see the condition of the orbiter postlaunch and in orbit; and improvements to scheduling and NASA organization. Even with all the efforts put into preparing the shuttle for a safe return to flight, foam falling from the shuttle at liftoff continued to be problematic, and new problems regarding thin gap-fillers between the tiles surfaced as well. At the final flight readiness review for the orbiter *Discovery* on June 17, 2006, NASA's top safety officer and chief engineer both voted not to launch based on continuing safety concerns. Administrator Mike Griffin overruled them, and *Discovery* was launched on July 4, 2006. The crew spent considerable time in orbit making sure there was no damage from foam that fell on liftoff that would preclude a safe reentry, and making sure that a leaky power unit would not impair the hydraulic systems used for steering and braking during reentry. They returned safely on July 17, but it was not a picture-perfect flight. Astronauts want to fly—that is why they become astronauts—and they accept a certain level of risk. But concerns over risks taken due to fulfilling timetables or political commitments are legitimate.

There are pragmatic implications for Florida aerospace workers in shutting down the shuttle programs, which could also affect safety. At a November 2004 news conference, O'Keefe stated that he knew of no definite plans for workforce reductions at Kennedy Space Center.[34] The same month, the *Orlando Sentinel* reported that 100 shuttle jobs would be phased out there: "the official denials are literally accurate, despite repeated references in NASA's own internal documents to 'layoffs.' Most of the job cuts will be accomplished through attrition, retirement, and leaving slots vacant—not handing workers pink slips. The bottom line, however, is that fewer people will be preparing the shuttle fleet for its return to flight. Those remaining may work longer hours at a critical time."[35] While short-term savings accrue, delayed schedules end up costing money in the long run, and they increase safety risks. Delayed schedules have also raised concerns about what happens if the schedules are not met.

Senator Bill Nelson (D-Fla.) quizzed NASA administrator Mike Griffin at a May 2005 Senate committee hearing[36] about NASA's "Plan B":

BILL NELSON: What happens if the CEV is not ready by 2010, when your plan is to scrap the shuttle?

MIKE GRIFFIN: It's my job to convince you that we will have a development plan for CEV that has it ready by 2010 or as soon thereafter as we can. We will say the date and we will try to hold to it, and I will try to convince you as we go through the next few years together, that we are holding to that plan. In part, the definition of what the CEV is needs to be done with the constraint that it be buildable, that it be an executable program and can be fielded shortly after the shuttle's retirement.

NELSON: And in that plan would be the plan for the orderly transition from one to the other?

GRIFFIN: Exactly. Yes, sir.

NELSON: Of course at that point we assume that the space station will be up in full bloom and running with a complete complement of astronauts doing research. Therein is another reason why we need to follow on vehicle ready, so that there's not this hiatus. Given the fact of our experience back in 1975, in the last Apollo, it was Apollo Soyuz, we thought we were going to fly the space shuttle about 1978 and it didn't fly until 1981. A part of that workforce was effectively utilized so that they didn't have to lay them off, and that corporate memory was all there. What are your plans, what is your thinking about, if there did occur this hiatus of how you would keep that team together?

GRIFFIN: Senator, I lived through that period as a working engineer and remember it well, and it's not one of my more pleasant memories from my, you know, 35 years in the space business. So one of the reasons I so strongly support the concerns which have been expressed about minimizing that gap in human space flight capability is that, I, frankly, don't want to live through that experience twice. Our primary contractor in space shuttle, space flight launch operations, launch appropriations is, of course, United Space Alliance. We work with that contractor every day, every week. We have very close ties with them. We're pleased with the work. It is our goal going forward, in developing a transition plan from shuttle to CEV, to be hand in glove with our USA contractor to affect the most orderly transition that we can. As I said briefly in my opening statement, we have studied lessons learned from the Titan 4 transition. We have gone back and studied lessons learned from the, you know, Saturn Apollo to shuttle transitions. Some of those lessons are good ones and some are things to be avoided. We are paying attention. There are two basic issues. Any launch system—I referred earlier to the fact, the known fact that it takes about $4.5 billion to keep the shuttle going, whether you fly any flights or not. The new system must have lower fixed costs or the United States will not have effected any improvement. Lower fixed costs means a smaller workforce in the sense of a standing army. So we want to shift money from what it takes merely to launch payloads into more exciting and new things that we want to do. So some workforce will be transitioned going forward and other elements of the workforce must be transitioned to new activities. Otherwise we'd do nothing new. And that program must be managed as carefully and as much of a forward-looking sense as we possibly can and that is our every intention.

NELSON: And an additional computation here is that you will have a full-up robust, internationally participated in, space station that you want to utilize. And suddenly if you stop the shuttle and you don't have the follow-on vehicle to service that space station, you can't use all of that investment of multiple tens of billions of dollars up in the heavens. For example, what would we use as a life boat? I guess we'd have to use the Soyuz. Well then, therefore, you got to drop the level of the crew, so you're not using that superstructure up there that we've invested so much in. What's your thinking there?

GRIFFIN: I cannot but agree with you. If we have a station, we need to be able to use it, and I am, I don't know how to say in enough different ways that I am convinced that your concern about minimizes such gaps in access on target, and we at NASA are working to eliminate that and to provide a credible plan for doing so.

NELSON: Well, I commend you on what you've already done, which is accelerate the CEV, and see if that is doable, but then that begs my next statement, which is, if it can't be accelerated, then maybe it's in the interest of the United States to extend the shuttle.

GRIFFIN: Sir, if we do that, then we face the circular problem. I have a hole in my gas tank, but the money I have to spend buying the gasoline prevents me from paying the mechanic to fix the tank. I've got to retire the shuttle in order to have the money to do the things that you and I both want to do.

NELSON: Understandably, but if the development of the CEV does not occur, in the expeditious way that you hope and that you're giving leadership to, and we commend you for that, you got to have a plan B. Otherwise we're going to waste all that asset up there. Just a concluding thought here, Madam Chairman. In a previous hearing we had brand new testimony about the promise, for example, that I did not know, of the experiments that are going on, on protein crystal drugs, experiments that were made 20 years ago and of which there was some question of whether or not it was financially feasible to do that on orbit, as opposed to on Earth. But now that we're seeing new promise, as was the testimony, I still have not received those answers, and I would like the statement to NASA is, please get those answers back to me. But if there is that promise of medical breakthroughs on such things like protein crystal growth on orbit, then that's just all the more reason why we need to keep that international space station functioning. Thank you.

The Hubble Space Telescope is also affected by the new NASA reorientation. As perhaps the most successful space observatory ever built, producing one astronomical revelation after another, it has a proven record of success. O'Keefe was willing to make the tough choice to allow Hubble an early death rather than using a shuttle mission to replace the spacecraft's batteries, which wear out in 2007. Others took a "not-so-fast" attitude. A twenty-one-member panel of the National Academy of Sciences issued a report in late 2004 urging O'Keefe to reconsider his decision. Sherwood Boehlert (R-N.Y.), chairman of the House Science Committee, and Senator Barbara Mikulski (D-Md.), representing the Goddard Space Flight Center housing the Hubble program, are among the bipartisan members of Congress urging a Hubble reprieve. Griffin promised when he took office to reconsider a shuttle mission to save Hubble, and as of July 2005, before the shuttle flights were suspended, NASA engineers were preparing to use the shuttle for a Hubble servicing mission as early as 2007.[37] The likelihood of that occurring sometime increased with $270 million included in NASA fiscal-year 2006 budget for a Hubble servicing mission.

But nothing is firm yet regarding Hubble's future. In September 2005, Griffin told Congress that "not one thin dime" would be taken from space science programs to pay for the exploration plan. Then, in February 2006, the decision not to take money from space science was reversed. In another slaughter of the innocents, NASA's 2007 budget included cuts of $3 billion to space science programs. Included in those program targeted were international programs. The Stratospheric Observatory for Infrared Astronomy (SOFIA) is a joint astronomy project between NASA and the German Aerospace Center. With the program 85 percent complete, and more than $500 million in taxpayer dollars invested in it—plus another $100 million from Germany—U.S. funding was cut for fiscal year 2007, mere months before the first scheduled test flight. The money saved from SOFIA and other programs was earmarked to cover the costs of the ISS and to get the shuttle back in flight for at least the sixteen space shuttle missions needed to complete the station. Pressure from the German partners resulted in NASA agreeing in June 2006 to "review" the decision to cut the program.

A multitude of considerations cloud future plans, made even more complicated because of the tenuousness of the presidential commitment to the vision, present and future. Curiously, when Congress voted to give NASA almost everything the president asked for, there was no reaction from those who had pushed for it, in either Congress or the White House. Legitimate speculation said that nobody wanted to draw attention to funding going for a program without widespread public support. Will the next president spend political capital to fulfill Bush's manned space exploration program, which the public sees as expendable? President Lyndon B. Johnson followed through on Kennedy's Apollo program, but it was, again, linked to a strategic goal, not just exploration. The same is not the case for the Bush vision.

The question of vision rationale has been puzzling. Exploration is the official reason for the Bush space vision, but while admirable, it will not sell in the long term. Without the requisite commitment to elevate it to a sustainable level, announcing a new manned space vision becomes little more than a good sound bite. Pushing the bill for it to future administrations certainly avoids realistic cost estimates, which have killed exploration programs in the past. But it also does not offer a politically viable long-term plan. That has evoked more cynical speculation that the vision announcement was politically motivated to generate votes before the 2004 election. At a conference near Kennedy Space Center before the election, it was certainly clear that workers at Kennedy Space Center believed that Bush had come to their rescue. But somewhere between altruism and cynicism there are two considerations. The Bush administration wanted to establish and embed as many technology development programs as possible that were potentially relevant to maintaining military space superiority and establishing space dominance, its key space priority. Also, the Chinese manned launch in October 2003 generated enough of an impression that there was a "space race," which the United States was losing, that a response seemed necessary. Likely, all these reasons played into the announcement.

On October 4, 2004, the forty-seventh anniversary of *Sputnik*'s flight, Burt Rutan's *SpaceShipOne* won the $10 million X Prize, given to the first privately funded manned spacecraft to reach space twice in one week. That event, followed in December by Congress passing the Commercial Space Launch Amendments Act of 2004, to allow space tourism, might portend a turn of events in space exploration. If wealthy thrill-seekers will pay large amounts of money for short rides into space, to experience zero-gravity conditions and take great pictures of the earth, private industry will continue to spend its own money for the technology to extend those trips, potentially including space hotels and excursions to the moon. That model of development mirrors that of automobiles, airplanes, and computers rather than the government-pays-all anomaly model experienced to date with space. It is a long-awaited and needed break with the old space development paradigm, and it will likely push technology development in ways that the government has been unable to do. Entrepreneurial development does not, however, replace the need for government initiatives.

In an era of unprecedented budget deficits, and as the costs of the war in Iraq grow with no end in sight, why should America care about manned space? Unfortunately for space buffs, exploration has never been enough, and it will not be enough now. Manned space activity yields benefits in the form of jobs, education, technology development, and prestige, but none of those is enough either. What manned spaceflight offers is a soft-power strategic alternative to counterbalance some, though not all, of the international fears generated by U.S. military ambitions in space. It could be a start toward ratcheting down some of the international (and likely domestic, if understood) concerns about U.S. military space ambitions that fuel the security dilemma discussed in chapter 1, and toward building relationships to alleviate U.S. concerns driving the current view that only technology and military strength provide security for the United States and its space assets. But it would have to be done right, and with political commitment equal to that afforded a security program, not a science program. That, unfortunately, does not describe the Bush space vision.

During the Apollo program, the world looked up to the United States not in fear but in admiration and respect. Admiration and respect can buy as much, if not more, security as a rifle can. Space offers opportunities to develop assets to further U.S. hard power and to cooperate with other countries toward generating soft power. Additionally, space cooperation offers the potential for the United States to influence technology transfer issues by moving to controlled sharing rather than futile attempts at denial. Public diplomacy is part of security, and maintaining U.S. leadership in manned spaceflight offers a highly visible and pragmatic diplomacy opportunity.

Fighting terrorism and dealing with the challenges of the twenty-first century require more than hard power; hard power is a necessary, but not sufficient, aspect of maintaining America's safety and preeminence. Other nations need to see the United States as the leader of choice to follow, rather than as a bully. In this regard, manned spaceflight offers an opportunity to be a leader, control more of the technology that the United States considers sensitive, and cooperate with other countries to develop the technology they deem to be essential to function and thrive in the globalized world.

As long as the U.S. manned space vision is purely about exploration, its future is uncertain. As long as its future is uncertain, so too is American leadership in manned space. It is ours to lose. Other countries, particularly Russia and China, have manned space capabilities, and China offers another option to countries that already have experienced partnerships with the United States. Letting go of its leadership in manned space might be tolerable under some circumstances, but not now. Currently, the United States is considering forfeiting an area that has long yielded soft power when soft power is most needed. At the same time, the United States is building its military space capabilities in ways perceived as threatening to friends and foes alike. The perception of threat must be countered symbolically, through cooperative manned space efforts. But the United States must also consider what its long-term military space objectives really are: Are they to maintain the heavens for exclusive U.S. use, or by U.S. permission only? The potential consequences of building space weapons demand thorough investigation; meanwhile, U.S. attempts to militarize space continue.

FOUR

The Militarization of Space

> If you look at the services, the future warfighting concepts, they are unexecutable without space capabilities.
>
> —Lieutenant General Eugene Santarelli, United States Air Force (Retired)

The U.S. military demonstrated its powerful combat abilities during the first Gulf War, the wars in Bosnia and Afghanistan, and the twenty-one-day march to Baghdad in April 2003. At more than $439 billion for fiscal year 2007, the U.S. defense budget is larger than the combined defense budgets of the next twenty top defense-spending countries. The military has replaced what is known as threat-based planning—planning for specific military threats from clearly identifiable enemies—with capabilities-based planning. The latter is intended to provide the capacity to defeat any conceivable attack, any time, anywhere. In effect, it prepares the United States to develop and maintain permanent military supremacy.[1] The second element currently driving Pentagon future force planning is transformation. Beginning with a report by the National Defense Panel, which was established to report in concert with the 1997 Quadrennial Defense Review, and elevated to priority status by Secretary of Defense Donald Rumsfeld, "transformation" is defined in a number of ways. In essence, it strives for lighter, faster, more agile, yet also more lethal combat forces. Technology, including space technology, plays a large role.

Space is embedded in the U.S. military's way of doing business and plays a large part in its success. In simple terms, space technology provides "see it," "state it," and "stop it" capabilities for the United States far beyond those of any other military. "See it" capabilities are those intended to provide, in military jargon, "universal situational awareness," the power to cut through the fog of war and obtain an edge over an opponent by having the most accurate information about the battle environment. "State it" capabilities are those involving command and control, with communication as the key. Knowing what is going on and being able to convey that to troops at the front line for effective use are two very different capabilities. "Stop it" capabilities are just what they sound like, ranging from nonlethal actions to precision-guided munitions.

The militarization of space is not a new phenomenon; developing space capabilities dates back to World War II and, in the United States, predates NASA. Military space activities are, however, less visible than civilian space activities. The increased incorporation of space-related technologies and capabilities into military operations until recently has occurred, to a large extent, as part of incremental modernization efforts. Moving from landline-based communications to satellite-based mobile communications has evolved in the military much as it has in the population at large. That evolution required initially developing and continually improving unique capabilities, such as space launch (hereinafter, "lift"). The U.S. military's increased use of space, especially compared to other countries, created technological dependencies, resulting in the perceived necessity to assume the zero-sum perspective about space discussed in chapter 1.

The Military Uses of Space

The four official national security space missions are space support, force enhancement, space control, and force application.[2] Until recently, the use of space assets has focused on the first two as force multipliers, in the form of technology to increase the potential of success for traditional forces. Transformation has accelerated the momentum of these efforts. There are, however, clear indications of an implicit drive toward force application through space control. The line between space control and force applications has been increasingly blurred since the 2004 issuance of the Air Force's doctrine, outlined in *Counterspace Operations*. This chapter provides an overview of the space missions generally, as well as a focused examination of space support and force enhancement, and the relationship between transformation efforts and space. Additionally, a discussion of the difficulty of deciphering U.S. intentions in space is offered as a prelude to a focused consideration of space control, including missile defense, and force application in chapter 5.

Space Support

Space support operations are the relatively unsexy but essential capabilities required to be a spacefaring country. They include activities such as launching and deploying space vehicles, maintaining and sustaining space vehicles while in space, and recovering space vehicles if required. The timing error mistakenly sent to GPS satellites in 1996 (discussed in chapter 2) was part of the space support mission. Placing satellites in orbit is also part of the space support mission.

Unquestionably, the most important issue in the space support mission has been and remains lift. Relying on launchers that evolved from weapons systems to being the workhorses of the launcher fleet has been the basic problem. Since the launchers were developed as weapons, their primary goal was to effectively deliver a warhead payload. As a weapon in a confrontation for potential national survival against the Soviet Union, the cost of launching the warhead was, understandably, not much of a concern. But as a launcher intended to place a satellite in orbit, cost is a large consideration.

The three basic factors relevant to discussions about launch are reliability, cost, and time. Reliability is most important because, as stated earlier, there is no market for a launcher that might or might not work—not with million- and even billion-dollar payloads on top. There is, however, sometimes room for trade-offs between cost and time. When a launcher lifts a satellite payload into orbit, the cost of the launch itself is involved, plus the cost of the payload, plus the cost of insurance on both the launch and the payload. Time refers to how long a customer has to wait to launch its payload after it has signed a contract. It can sometimes cost customers more to store their payloads while waiting for a launch date, or in potential lost revenue, than to pay a higher price to a vendor that can provide faster launch services. All these factors weigh into customer decisions.

Retired general Tom Moorman Jr., former commander of Air Force Space Command, is attributed with the claim that "the Earth is covered by two-thirds water and one-third space studies, most of them chaired by me." After the *Challenger* accident in 1986, a plethora of government and private studies were generated to determine how to push space activity out of the doldrums that consumed it.[3] All examined, stated, and restated the need to address the issue of earth-to-space transportation as the number-one priority. Fix lift, they said.

Lift was considered to be broken because the average cost per pound to launch payloads into orbit was so high that nobody but governments could afford to go to space. The limited market restrained the willingness of the private sector to invest in new technologies needed for the sector to develop more normally. Additionally, "access" was anything but easy. Rockets were built and mated to specific payloads, often taking inordinate amounts of time. The need to move toward standardization, whereby payloads could be adapted and placed on interchangeable launchers faster and more easily, was recognized as well.

There are two types of lift: expendable launch vehicles (ELVs), so named because they can be used only once, and reusable launch vehicles (RLVs). The space shuttle is currently the only operational RLV (and it is only partially reusable), and RLVs have generally been considered the purview of NASA.[4] Prominent among the ELVs are the Atlas, Titan, and Delta, all originally derived from ballistic missile technology.[5] Other ELVs have been developed more recently by private companies, usually for smaller payloads: Pegasus and Taurus, by Orbital Sciences Corporation, and Athena by Lockheed Martin. Differentiations between rocket capabilities, beyond expendability or reusability, are made according to how much weight the rocket can lift (payload capability) and at what altitude it can put its payload (table 4.1). A general rule of thumb is that the heavier the payload and the higher the orbit, the more expensive the launch.

At the time of the post-*Challenger* studies in the late 1980s, the average cost per pound of launching a payload into GEO, where many large spacecraft reside, using a Delta, Atlas, or Titan launcher was over $20,000. Launching a 12,000-pound payload on a Titan 4 cost in the hundreds of millions. Waiting for a turn to use limited launch pads could drive costs up further, which is especially problematic because commercial payloads and military payloads often rely on the same launch facilities in the United States, and military payloads take priority. When specific payloads are mated to specific launchers, a problem with one means the other has to be stored. Beyond costs of potentially thousands of dollars a day, storage can create technical problems as well, which then have to be corrected at additional cost. Between 2001 and 2003, a Defense Meteorological Satellite Program (DMSP) satellite had to have its propulsion system replaced after an acid residue ruined the first system during one of the launch delays that plagued the satellite for thirty-three months from its first scheduled launch date until it was finally launched.[6] For commercial satellites, expensive insurance is also a consideration.

Global satellite owners, satellite manufacturers, launch service providers, insurance brokers, underwriters, financial institutions, reinsurers, and government agents cooperate to provide a coordinated insurance package for any given commercial satellite launch. There are many types of insurance, including

Table 4.1 Exemplary Rocket Payload Capabilities (in pounds)

Expendable Launch Vehicle	Payload Capability to Low Earth Orbit (LEO)	Payload Capability to Geostationary Orbit (GEO)	Estimated Cost of Launch* (millions of dollars)
Atlas 5 (EELV)	20,000	10,900	$138–192*
Delta 4 (EELV)	50,000	9,200–14,000 (to geostationary transfer orbit)	$138–254†
Titan 4	49,000	12,700	$500
Pegasus XL	900	LEO only	$13
Space shuttle	55,000 (of cargo)	LEO only	$300–500
Athena 2	4,390	1,300	$24

Note: For more detailed information on rocket systems and costs, see Futron Corporation, "Space Transportation Costs: Trends in Cost per Pound to Orbit, 1990–2000," September 6, 2002, available at www.futron.com/pdf/FutronLaunchCostWP.pdf; and Marcia Smith, *Space Launch Vehicles: Government Activities, Commercial Competition, and Satellite Exports*, issue brief IB93062, prepared by Congressional Research Service, October 6, 2003.

*Depending on whether the medium-powered or intermediate-size version of the launcher family is used.

†Depending on whether the medium-powered, intermediate-size, or heavy-lift version of the launcher family is used.

liability if a failed satellite should crash to earth, insurance to cover the loss of a satellite in case of a failed launch, and insurance to cover the satellite in case it simply does not work once in orbit. The average cost of launch insurance is around 15 to 20 percent of the total launch price, but it can rise as high as 30 percent or more if launcher reliability is questioned. If that happens, the launch provider can find itself priced out of the market.[7]

The post-*Challenger* reports acknowledged all these issues. Several false starts toward fixing the lift problem resulted in millions spent to no avail. Knowing what had to be done and being able to do it seemed impossible, because nobody could explain why a new launcher was needed. What would be launched with a new launcher that could not be launched with available launchers? Suggestions to the government and private industry that they invest millions, more realistically billions, to build a new launch vehicle to offer affordable, reliable, and easy access to space fell on deaf ears. Everything that needed to be launched was being launched, just at a high cost. Many commercial space ventures were stifled by high launch costs, except for communications, in which costs could be passed along to large consumer markets dependent on communication satellites. The private sector was looking for government investment; the government was looking for the private sector to use its own money to build a new launcher. But "build it and they will come" was not viewed as a wise commercial investment approach.

For the military, the lift issue had other dimensions as well. During the Gulf War, Chuck Horner, an Air Force general and a commander of the allied air forces, was provided satellite imagery for targeting and poststrike assessments. A fighter pilot by profession, he was impressed with the imagery and sought more. With the battle raging around him, he called the Air Force Space Command and said he wanted another satellite. The person he reached was thrilled that the value of the Space Command's input was being recognized. "When can I expect imagery from the new satellite to be available?" Horner asked. "Well sir," was the eager response, "we should be able to get a launch date in about two years."[8]

Finally, looking at the long-term military space budget and seeing just how much of that budget would be devoured by launch costs, the military and Congress made a move. In December 1993, Congress formally tasked the Department of Defense to develop a space launch modernization plan (SLMP), and General Tom Moorman Jr. was commissioned to lead the effort. His task included one political parameter. During this time, the United States was trying desperately to keep the Russian space program afloat, rather than watching Russian launch (missile) experts flood the global job market. So, President Clinton wanted the Russians included in whatever was built. Moorman's SLMP team developed four modernization options: sustaining existing launch systems; improving—or, in the parlance, "evolving"—current expendable launch systems; developing a new expendable launch system; or developing a new reusable launch system. Evolving current ELVs was selected as the only economically viable choice, and the Evolved Expendable Launch Vehicle (EELV) program, including use of Russian rocket engines, was born.

The private sector's interest in the EELV was based on strong forecasts for launching commercial communication satellites during the mid-1990s. Growth was expected through the early twenty-first century. There was money to be made from commercial launches, intended to offset investments the government expected them to make in developing the EELV. Government–private sector partnerships were envisioned to get EELV off the ground. The intent was to build a family of launchers, modernize existing launchers, and provide affordable launch options depending on spacecraft size, weight, and destination requirements. The mission statement for the EELV program read, "Partner with industry to develop a national launch capability that satisfies both government and commercial payload requirements and reduces the cost of space launch by at least 25 percent."[9] The EELV received its first funding, $30 million, in fiscal year 1995 as a DOD program.[10] That was a down payment on what was anticipated as a $17 billion program.

The initial intent was for one company to be selected to build the EELV, but government concern about the need to keep more than one launch builder healthy and viable resulted in both Lockheed Martin and McDonnell Douglas (later bought by Boeing) receiving pre-engineering and manufacturing development contracts worth $60 million each in 1996. Lockheed Martin, Boeing, and the Air Force were each to contribute $1 billion to developing the Atlas 5 and Delta 4 as the core of the EELV plan. Atlas 5 uses the best of the Atlas and Titan programs, as well as incorporating a highly proven Russian RD-180 rocket engine, to satisfy the foreign policy aspect of the EELV launch strategy. The Delta 4 combined new and mature technologies, including the RS-68 engine with no previous flight history, to launch virtually any medium or heavy payload into space.[11]

An Atlas 5 and a Delta 4 were successful launched by 2002, but not without more government funding than initially budgeted.[12] At one point, Congress took to referring to the EELV as the "extremely expensive launch vehicle." This was between 1998 and 2002, when the DOD refused to provide Congress with key data on program costs and goals.[13] The DOD feared that revealing the extent of program cost overruns would jeopardize the program's continuation. The government's contribution to the EELV program jumped during that time, from an expected $17 billion lifetime cost to a still-growing $32 billion, more than the cost overruns for the International Space Station that attracted considerable attention and calls for cancellation. Part of the ballooning cost was likely because of the political need to underestimate costs to get programs approved (discussed regarding the space shuttle and the ISS in chapter 3) and a problem for virtually every military space program. Military programs underestimate their costs so commonly that Congress is almost numb to it. The other unfortunate circumstance specific to this case was a drop in commercial launch projections in 1999. When that occurred, it became clear to Boeing and Lockheed Martin that the price that they had charged for the military launches was too low and would have to be increased.[14] Technically, however, launch-cost savings sought over older systems were achieved, as the EELV cost per pound to orbit initially averaged about $10,000, about half of what it had been.

As much or more of the launch cost savings achieved by the EELV was through government acquisition law reform as through new technology. In 1999, members of the EELV team from the Air Force and the Aerospace Corporation received the DOD's highest award for sound acquisition practices. Why savings were garnered through acquisition rather than using new technology illustrates larger issues within the aerospace field. First, government acquisition laws have long been convoluted beyond belief; those who bureaucratically perpetuate the system have no incentive to change unless prodded externally. The system does not encourage government purchases for efficiency or maximum effectiveness; it favors the status quo and suppliers that know how to work the system. Second, modifying old technology and gaining ground from streamlining acquisitions, rather than using new and improved technology, are too often the only viable choices. Incorporating new technology into a space system, or using totally new technology, is not only technically risky, but very expensive, sometimes prohibitively so, because of the need to test the new equipment. Developers can use old technology that has already been tested, use new technology and potentially bust their budgets on testing, or forgo testing and risk failure. A new airplane engine will undergo hundreds, if not thousands, of hours of ground testing before it is flown. Because of being over-budget and tightly constrained (risking cancellation), the space shuttle main engine was ground-tested for less than one hour before the first shuttle launch.

Meanwhile, the military has long coveted manned space capability. Ever since John F. Kennedy gave the manned mission to the upstart civilian NASA, largely because he and Secretary of Defense Robert McNamara tired of the prevailing interservice mission rivalry that resulted in more bickering than brilliance, the military has tried back-door efforts to get a foothold in manned space programs. The Air Force Manned Orbiting Laboratory (MOL) and shuttle precursor Dyna-Soar programs in the 1960s, complete with six Dyna-Soar astronauts in 1962, evidence that desire. McNamara canceled both. Plans for a military space plane have surfaced more than once, too, though Air Force leadership's reluctance to spend development funds on space rather than a new fighter airplane (the F-22) has kept it on the drawing board.[15]

To be a space power, rather than a space services consumer, a country must be able to get to and maneuver in space. That means that until lift is easy and affordable, it will continue to be an issue. Military needs can be met with current capabilities, but at a substantial cost. The desire to rapidly deploy space weapons, if pursued, could provide the impetus for a changed attitude about space transportation. In fact, it could provide the motivation to develop the space plane long sought by the military.[16]

The Air Force is already funding the development of a hypersonic munitions delivery system. Originally known as the Common Aerospace Vehicle (CAV), the program was a combined Air Force and Defense Advanced Research Agency (DARPA) effort, with technical assistance from NASA, evidencing the indirect ways NASA contributes to military efforts. Part of a program dubbed Falcon (Force Application and Launch from Continental United States), it also includes a small launch vehicle to carry the CAV to an orbit altitude, from where it would coast to its "pierce point" location—that is, the point and velocity at which the vehicle enters the atmosphere. The Air Force originally intended to field the CAV by 2011, allowing the United States to attack any target on the globe within twelve hours of an order to do so.

In 2004, however, Congress stepped in. Concerned with clear Air Force zeal to move ahead with space weapons programs, lawmakers directed the Air Force not to engage Falcon in any weapons-related work during fiscal year 2005, and halved the funding for the CAV.[17] Subsequently, the Air Force said that the CAV had been canceled. In reality, however, it was simply renamed the Hypersonic Technology Vehicle[18] and is scheduled for its first test flight in September 2007.

Force Enhancement

Force enhancement capabilities are those that, when added to and employed by a combat force, significantly increase the combat potential of that force, and thus enhance the probability of a successful mission. The process described earlier of attaching GPS packages to bombs to create joint direct attack munitions is an example of force enhancement. Force enhancement also includes a wide range of capabilities that help decision makers and troops assess a situation and communicate with one another. Other examples include the use of remotely sensed imagery, weather satellites, communication satellites, and the various components of signals intelligence (SIGINT). All these help the military break through the fog of war and have been increasingly incorporated into U.S. military operations since the 1991 Gulf War.

In essence, SIGINT involves intercepting all types of electronic emissions to gather information on a target, including listening in on enemy radio conversations, which may require code-breaking or other cryptological expertise. Even radio signals that do not carry voice messages, such as radar signals, can be useful: detecting radar signals can supply valuable information about enemy positions and strength. Collecting information on a target's electronic emanations to determine its countermeasure (defensive) capabilities, including jamming and electronic deception, are also important. Some data can be gathered through airborne platforms, ships, or submarines, but increasingly, data is gathered by satellites.

Information is clearly important. Having the most and the most quickly available information is a nearly insurmountable advantage in warfare. To quote the ancient Chinese strategist Sun Tzu, "What enables the wise sovereign and the good general to strike and conquer, and achieve things beyond the reach of ordinary men, is foreknowledge and the decision from it." At this point, the United States both has the highest capabilities and is the most dependent on those capabilities. That makes protecting those assets imperative.

Space Control

While "space" has been an Air Force mission, rather than just a place, since 1988, space control has been officially included as an Air Force space mission only since 1996, and its nebulous definition initially resulted in its being a less actively pursued space mission than space support or force enhancement. With the steady infusion of space capabilities into military operations, however, that is changing. Space control basically means ensuring that the United States can reap the advantages offered from space, and adversaries cannot. Counterspace operations, as defined in Air Force Doctrine Document 2-2.1, supports the space control mission because it refers to what can be done to protect space assets.

When considering how to defend U.S. space systems, or attack an adversary's systems, it is important to recognize the three segments involved: the ground segment, controlling the system operations; the space segment; and the electromagnetic links connecting the ground and space segments. The ground segment is the most vulnerable, if for no other reason than the relative ease of physical access. Ground segment security to protect against attack includes both physical protection such as facilities hardening, which can literally mean using concrete bunkers, and personnel and information security, such as firewalls and encryption. Additionally, dispersed mobile ground stations can provide system backup. There has been, however, rising concern about possible attacks on other parts of the system. Perhaps most infamously, the 2001 Space Commission sounded the alarm about a possible "Space Pearl Harbor." If one accepts the premise that the United States is vulnerable to such an attack, then it can be argued that there is an obligation to defend against it.

What would a space Pearl Harbor look like? It could involve the damaging or destruction of one or more reconnaissance satellites, communication systems, weather or earth observation systems, navigation satellites, or satellites for detecting missile launches. Interfering with a reconnaissance satellite would diminish the military's "see it" capabilities—its knowledge about enemy strength and capabilities—and consequently reduce effective planning ability, accurate targeting, and battle damage assessment. Weather and earth observation systems allow more effective operational planning. Interfering with navigation satellites complicates troop, aircraft, and ship movements; can impede the use of precision-guided munitions; and can encumber missile launch detection satellites, increasing the possibility of a surprise attack. Damaging communication systems interrupts command and control among troops—the military's "state it" capabilities. The question then becomes what can be done to prevent or counter these potential events. The answer is not inherently space weapons.

Generally speaking, a variety of technical and nontechnical means can protect the space and electromagnetic link segments of a space system.[19] A number of passive systems are already being utilized to various degrees. Hardening, for example, involves protecting satellite components from harm, including from directed energy, laser, or microwave weapons, by using hardware such as filters and shutters that cover optics. These techniques are also useful to protect satellite components from damage during naturally occurring events, such as meteor showers. Shielding, another protective technique, includes using metal shields and resistant paints to protect satellites from electromagnetic pulses (EMPs), generated either naturally by solar storms or by nuclear blasts or other weapons. Circuit-breaker-like devices in satellites can switch off nonessential components to prevent or minimize possible damage from an EMP.

Denial and deception, just as they sound, refer to preventing an adversary from learning about satellites, using techniques that reduce electro-optical and electromagnetic satellite signatures with thermal blankets and energy-absorbing materials. Deception methods intentionally mislead adversaries about space systems. Another way is referred to as stealth and cloaking. Energy reflected by spacecraft can be detected by radar, infrared, or acoustic sensors; minimizing energy reflection and camouflaging spacecraft can help spacecraft avoid detection. In the future, "adaptive skins" that change molecular characteristics and deflect or absorb incoming energy may become technically feasible. Using decoy satellites has also been suggested to confuse adversaries. These are all protections incorporated into a spacecraft design or involve hardware and, subsequently, increase the cost of the spacecraft. The additional weight from such designs can add to the launch cost.

Other protections involve techniques that do not use hardware. Simply moving away from adversarial spacecraft, referred to as maneuvering,[20] is one way. Providing backups for critical spacecraft and system components, known as redundancy and reconstitution, also decreases vulnerability. The GPS always has spare satellites in orbit. The idea that the United States must be able to reconstitute space assets faster than any other country is part of a doctrine called responsive space.[21] By neither clustering satellites nor making their location predictable (possible for some satellites, but not others) through dispersion, it becomes harder for adversaries to locate and target valuable assets.

Offensive defensive capabilities, a third category of "active" protections, are all in various stages of R&D. Spacecraft "bodyguards" are envisioned by the Air Force as an integrated fleet of networked small satellites designed to detect adversarial satellites and possibly to "negate"[22]—that is, destroy—their threat. These small satellites pose issues of their own, however, which are discussed in chapters 5 and 8 in conjunction with space weapons and Chinese space activities. Directed energy weapons project intense energy to disable or destroy satellites. Kinetic energy weapons[23] use high-velocity projectiles to destroy targets—basically, slamming one object into another, with no explosive warhead involved. While both Congress and the military emphasize that the United States does not have a space weapons program, the United States does have programs in all of these areas, many under the pretext of space control.

Clearly, a wide range of options are open within space control, some passive and some active, some technical and some nontechnical. Discussions about active systems inherently moves to space weapons, whether deployed from the ground or from space. Part of the issue with active systems is the inability to differentiate between the technology required for defensive and offensive use, a key factor exacerbating the security dilemma. That issue would be further complicated should directed or kinetic energy weapons be used in preemptive attacks. Use of nontechnical means, arms control, and "rules of the road" agreements have not been pursued; in fact, the trend has been away from those options.

Transformation

There is an expression that "only poets write strategy without a budget." The United States has the largest defense budget in the world, but it is relevant to question whether it has the economic and political wherewithal and the manpower to be the world's policeman, as a country with permanent military supremacy might be anticipated to do. Since the mid-1990s, transformation has been offered as the solution to that dilemma. It took center stage at the Pentagon under Secretary of Defense Donald Rumsfeld. Transformation of the military involves several components, but basically means making broad changes in both structure and doctrine to meet challenges from terrorists, potential near-peer competitors, asymmetric warfare, insurgencies, and information warfare in the twenty-first century. A large dose of advanced technologies to get more bang for each defense buck is included as well.[24] Transformational technologies are those that revolutionize the nature of warfare, such as airpower, rather than merely being part of evolutionary modernization. This is where space technology becomes important.

Rumsfeld presented his objectives for transformation in a speech at the National Defense University in Washington on January 31, 2002.[25] They include protecting the U.S. homeland and bases overseas; protecting and sustaining U.S. power around the world; denying sanctuaries to U.S. enemies anywhere on earth; protecting U.S. information networks, telecommunications, computer systems, and Internet links from attack; using information technology to allow different U.S. forces to communicate with one another to fight jointly; and maintaining unhindered access to space, while protecting U.S. space assets from attack. The importance of space is clear, through both the explicitly stated need to protect assets and the implicitly assumed use of space assets to enhance the capabilities of U.S. forces. The linkage between space and information technology alone makes space a key transformation tool.

A smaller and smarter military can theoretically overcome brute-force opponents, even when outnumbered, with the use of sensor-dependent networks. Sensors on the ground, in the air, and in space provide information key to modern warfare. Large parts of the Pentagon's transformation efforts rely on such networked systems. Images of high-tech ground, air, and space systems linked—schematically, by lightning bolts—to increase the multiplier potential of each individual system have been common in Pentagon literature and briefings for some time. The problem is that, according to the scientists and engineers tasked with creating these linked systems, each lightning bolt can potentially cost about $1 billion, as well as a great deal of time for development and testing.

Not to drive this point home too many times, but technology generally, and space technology specifically, is expensive. One military communication satellite program, called Milstar, cost an estimated $17 billion. For the same money, the military could have purchased two aircraft carriers, based on the next-generation aircraft carrier (CVN 21) price of $8 billion,[26] or seven B-2 stealth bombers, at $2.2 billion a plane. Meanwhile, Milstar is now scheduled to be replaced by advanced extremely high frequency (EHF) satellites. In 2001, when the contracts were awarded to Lockheed Martin, the EHF satellites were expected to cost $3.2 billion each. By 2004, that estimate had reached $5 billion, with another 50 percent increase considered not unlikely before program completion. High costs are the rule rather than the exception. Technology is also complicated and fallible. Once the actual invasion of Iraq started in 2003, breakdowns quickly surfaced. Information about the battlefield, such as troop locations, that was available from space assets at U.S. military headquarters in Qatar was not reaching the front lines. There were problems even with relatively simple systems for the movement of data, such as satellite images. Downloads took hours while battles raged. As many times as not, American troops found the enemy by running into them, as they did in World War II.[27] Technology is great, when it works.

A key requirement to transmitting information is bandwidth; the more bandwidth available, the more and more quickly the data can be transmitted. Many of the satellites that provide bandwidth are commercial, creating both opportunities and vulnerabilities. During an exercise regarding a potential U.S. military intervention supporting Taiwan, the notional adversary, China, managed to buy up surplus commercial bandwidth capabilities, leaving the U.S. military and its enormous bandwidth needs helpless.

The U.S. commanders directing the Iraqi war from Qatar and Kuwait had forty-two times the bandwidth available to them than had their counterparts in 1991. But, as stated previously, there were still considerable problems getting material from the commanders to the front lines. The Sixty-ninth Armor of the Third Infantry Division was at the very tip of the U.S. Army's final lunge north into Baghdad, taking a bridge as part of Objective Peach, a direct route to Saddam International Airport. Yet according to Lieutenant Colonel Ernest "Rock" Marcone, a battalion commander, they were almost devoid of information. "I would argue that I was the intelligence gathering device for my higher headquarters," he said.[28] Marcone could get no information on who or what was defending that bridge. Finally, he was told there was one Iraqi brigade at the bridge. There were actually three brigades, of twenty-five to thirty tanks, seventy to eighty armored personnel carriers, artillery, and between 5,000 and 10,000 Iraqi soldiers coming from three directions, facing a U.S. force of 1,000 soldiers with thirty tanks and fourteen Bradley fighting vehicles. The Iraqi contingent was a conventional force, which is exactly what space reconnaissance assets are designed to detect. Not only did space reconnaissance assets fail to provide information useful to Marcone, but communication assets provided him no information at all.

Operation Iraqi Freedom saw many technical successes as well, however. One was a vehicle-tracking system known as Blue Force Tracker. Using space-based GPS, troops were able to identify forces as friendly or not. A U.S. radar plane detected an Iraqi Republican Guard unit advancing on U.S. troops during a blinding sandstorm that raged from March 25 to 28, 2003. Bombers using satellite-guided munitions, able to work despite poor visibility, found enemy targets while distinguishing nearby American forces, an amazing technical success. However, there were not enough Blue Force devices to go around, or enough bandwidth in the field to support those that did make it to the front. On average, Army divisions got about 150 Blue Force systems per division, which meant that commanders and reconnaissance vehicles might have had them, but not individual vehicles. Further, satellite communications delivered Blue Force data, but competition for bandwidth inhibited the process.[29]

Clearly a prodigious force enhancer, technology is also a potential Achilles heel, because of both the dependence it creates and the ease with which an enemy can recognize that dependence. The reliance of the United States on technology, including space technology, presents an opportunity for an asymmetric challenge to U.S. military superiority, one that does not rely on equivalent military force. Moreover, an asymmetric challenge is the only challenge available to opponents. As Chinese analyst Wang Hucheng stated, "For countries that can never win a war with the United States by using the methods of tanks and planes, attacking an American space system may be an irresistible and most tempting choice." The quote is one of braggadocio—attempting to make the point that the United States can be beaten. It was pulled from an article entitled "The U.S. Military's 'Soft Ribs' and Strategic Weaknesses," originally printed in *Liaowang*, a decidedly anti-American publication and one that certainly represents the anti-U.S. perspective, though it was later reprinted by a Hong Kong news service.[30] But there is also an element of asymmetric truth being stated, much like the response attributed to India's then-chief of staff when reporters asked him what he had learned from observing the conflict in Iraq during the Gulf War. "Don't fight the Americans without nuclear weapons," he replied.[31] From the U.S. perspective, the answer to such challenges seems to be exclusively viewed as being "build more technology" ourselves and deny it to others. That creates a situation aptly described by retired admiral Eugene Carroll Jr.: "For 45 years of the Cold War we were in an arms race with the Soviet Union. Now it appears we're in an arms race with ourselves."[32]

In force-on-force conditions, the United States has proven itself to be unstoppable. Increasingly, however, urban and guerrilla warfare are becoming the norm, and success in those environments is less dependent on a technological edge. Flak jackets and armored personnel carriers can be more useful than missiles. Equally important, and largely neglected, is the question of whether transformation tools are the right ones for the problem, now and in the future. According to retired major general Robert Scales, maybe not. In a March 2005 article in *Armed Forces Journal*, Scales talks about attending a high-level seminar to discuss future threats to the nation, which focused almost exclusively on China.[33] Though the audience was agitated by the neglect given the Middle East, the DOD speakers were nonplussed. Apparently, Pentagon defense transformation efforts require a "peer competitor" worthy of our weaponry, and China has won the audition. As General Scales described the Pentagon approach, "We have the hammer. Now all we need is the right nail." His concern is that while defense planners must look to the future, including China's position, they need to do so without ignoring the realities of the present.

The present reality is that the United States is fighting an enemy that is largely "unplugged"—neither reliant on nor particularly thwarted by sophisticated technology. As Scales argues, "We need a new vision shaped by the realities of this new era—realities that we inherited rather than the one we created when transformation began in earnest four years ago."[34] American analysts sometimes cite the Chinese term *shashoujian* as the rationale for China being perceived as the future threat to the United States. The word first gained prominence, or perhaps more accurately, notoriety, after appearing in the report of the U.S.–China Economic and Security Review Commission in 2002.[35] While *shashoujian* can be translated a number of ways, including "assassin's mace" and "silver bullet," it is not a purely military term. Until recently, it was more often used in conjunction with economics and social relationships,[36] describing who has the advantage in getting a date with a pretty girl. In the United States, however, the conjecture has grown that it refers to clandestine, high-tech weaponry that China is developing and the United States must be prepared to thwart. The operational meaning of the term is debatable, but nevertheless, the quest for transformation capabilities to counter it seems unappeasable.

Deciphering U.S. Space Intentions

A September 2004 Air Force conference provided the venue for Peter B. Teets, undersecretary of the Air Force and director of the National Reconnaissance

Office, to present America's vision for space. "Even though we have superiority in many aspects of space capability, we don't have space dominance, and we don't have space supremacy," he said. "The fact is, we need to reach for that goal. It is the ultimate high ground."[37] This call to take space as the "ultimate high ground" is often heard, though the comparison with other high grounds is not without flaws. Further, what does "space dominance" really mean?

The short answer is that nobody truly knows what space dominance means, though there are many ideas and a great deal of activity, often outpacing answers to questions or even debate on the issues. Is the United States attempting to "lock out" other countries from pursuing space activities? Theresa Hitchens states that "that all nations have a right to space access including for military purposes and that all nations have the right to reaction (proportionally and consistent with international law) to protect and defend assets."[38] Would the United States support this sovereign principle for all countries, or only for itself? Is the United States intending to develop and deploy space capabilities to demonstrate that it does not accept even potential challenge, such as Galileo? Is the United States contemplating both space-to-space and space-to-earth weapons? Is there a significant difference? For the most part, these questions are not even being asked, let alone answered. Consideration of the long-term issues associated with military space activity has been overtaken by events and ambitions.

What would justify spending vast amounts of money to build a missile defense system that provides little defense capability, squeezing other countries out of space, and developing weapons that potentially turn near space into a shooting gallery? Policies, statements, and Sunday-morning talk shows featuring sound bites from Washington abound, but interpreting them to get a clear picture of U.S. intentions is every bit as hard as interpreting Chinese intentions, a source of great heartburn in Washington. Secretary of Defense Donald Rumsfeld has provided some insight:

> Our goal is not simply to fight and win wars, it is to try to prevent wars. To do so, we need to find ways to influence the decision-makers of potential adversaries, to deter them not only from existing weapons, but to the extent possible, try to dissuade them from building dangerous new capabilities in the first place. Just the existence of the U.S. Navy dissuades other from investing in competing navies—because it would truly cost a fortune and would not succeed in providing a margin of military advantage. We must develop new capabilities that merely by our possessing them will dissuade adversaries from trying to compete. For example, deployment of effective missile defenses may dissuade others from spending

> to obtain ballistic missiles when they cannot provide them what they want—the power to hold U.S. and allied cities hostage to nuclear blackmail. Hardening U.S. space systems, and building the capabilities to defend our space assets, could dissuade potential adversaries from developing small "killer satellites" to attack and cripple U.S. satellite networks.[39]

According to this statement, deterrence—preventing wars—can be achieved by dissuading potential adversaries from even trying to compete with the United States. By lowering the probability of a successful missile strike through missile defense, potential adversaries will be dissuaded from firing a missile at the United States in the first place. Also, apparently, if the United States has space weapons, then other countries will be dissuaded from building them or interfering with U.S. assets.

General Lance Lord, commander of the Air Force Space Command, speaking at hearings on the fiscal year 2006 space budget, expands further on U.S. intentions:

> We must prepare to face future threats today. My top priority is to ensure Space Superiority. This is at times a difficult concept to comprehend. We did not choose saber rattling words. We selected doctrinal terms; words we know are well understood in the Air Force and throughout DOD. The term Space Superiority is akin to Air Superiority. We would not dream of conducting air operations without first establishing Air Superiority. We are not trying to dominate, but we must protect and project our interests in the space medium.[40]

Leaving aside the contradiction between Teets's and Lord's statements regarding whether dominance is indeed sought, General Lord says that the United States seeks to both protect and project U.S. interests. The former could include both protecting terrestrial targets, justifying missile defense, and protecting space assets, justifying space weapons. The latter, however, adds a new twist to U.S. space missions.

The notion of power projection through space is evidenced in other documents as well. The National Defense Strategy (NDS) released in March 2005 states that space "enables us to project power anywhere in the world from secure bases of operation." Consequently, according to the NDS, the United States needs "to ensure our access to and use of space and to deny hostile exploitation of space to adversaries."[41] Projecting interests through, from, and in space entails missions that the U.S. public is largely unfamiliar with, beyond having ground-based missiles travel through space to reach a target. Additionally, power projection from space could include using space-based assets to prevent an adversary from acting; for example, using a laser to preemptively destroy a launch pad on earth, or a satellite in orbit.

This use of space assets for preemptive purposes extends other methods of power projection, such as aircraft carriers, long-range bombers, and ground-based missiles. It is part of the "preventive war" theory that itself extends the preemption doctrine that first came to prominence with the post–September 11 national security strategy and then the Iraq war and that has generated considerable discussion among security analysts about the future of war. That discussion focuses on not just ideological pros, but pragmatic cons. Is preemption intended as an exclusive option of the United States? Other countries likely will not see it that way. Since the United States began stressing its willingness to preemptively strike opponents, others, such as France, have as well. Whether the United States really wants to expand and export the preemption doctrine is questionable, though some would argue inevitable.[42] While it is still being debated, however, we are already building technology with the potential to preemptively strike targets from space.

When technology begins to outpace policy, "hedging" can be a useful alternative. Hedging is the notion of minimizing risk by counterbalancing one action against another.[43] In the case of space technology, hedging promotes R&D in areas not needed now, to allow rapid technology development and deployment if needed in the future. Funds can thus be used in areas more urgently required. Hedging avoids prematurely proceeding with technology development and deployment before policy implications are considered. It is a prudent option that the United States can easily undertake, the equivalent of buying an insurance policy without breaking the bank. It also allows for reasoned consideration rather than the cart leading the horse down a path not necessarily in its interests.

The statements by Rumsfeld, by Lord, and in the National Defense Strategy provide a basis from which to examine a number of critical premises behind the expansion of U.S. military space capabilities from an effects-based perspective. What are we trying to do, and will the actions achieve the intended goal? It appears that there are two general objectives. One is deterrence: preventing a missile strike against the United States by lowering the potential of success, and preventing wars by dissuading potential adversaries from trying to compete in building weapons systems. The other is space dominance: projecting U.S. interests in the space medium and protecting U.S. assets. Examining these goals more closely clarifies the issues for consideration.

Strategic deterrence is defined as preventing an adversary's aggression or coercion from threatening the vital interests of the United States or its national survival.[44] Deterrence is intended to convince adversaries not to take particular courses of action by decisively influencing their decision making. The idea is to affect an adversary's "strategic deterrence decision calculus," which involves three primary perceptions from the adversary's point of view: the benefits of a course of action, the costs of a course of action, and the consequences of restraint—that is, what happens if the adversary does not take the course of action. Similarly, there are three primary and sometimes overlapping tools to implement effective strategic deterrence: denying benefits, imposing costs, and inducing adversary restraint.

In the post–Cold War world, the assertion has been made that deterrence no longer works because of the changed geopolitical climate from the past. During the Cold War, countries threatening each other—specifically, the United States and the Soviet Union—could be assumed to act rationally when it came to risking their very survival, because the country was more important than the leader or an ideology; the benefits of nuclear weapons were denied by the accompanying cost being too great. This is the essence of the Cold War doctrine of mutually assured destruction. With threats increasingly coming from nonstate or rogue state actors, however, that premise becomes questionable. With missile defense as a defensive system, the United States is trying to induce adversary restraint by lowering the probability of success. But will it work, and is that the best option? The value of missile defense to deter a missile strike against the United States is technically questionable and strategically limited at best. The value of missile defense as an explicit part of space superiority or an implicit part of space dominance may better justify the money being spent. This gets into the idea of potentially using missile defense for more than simply defense against ICBMs, but rather as space weapons. Where is all the interest coming from?

Military advocates assert that any program that hints of space weapons gets scrutinized by Congress and assiduously pulled from the Pentagon budget. But the relationship between Congress, as the holder of the purse strings, and space weapons is complicated. Most members of Congress do not support the United States blatantly developing space weapons. Some members, primarily in the Senate, do fundamentally believe that space weapons, by some definition, are necessary to protect the United States, and these members are both adamant and effective in advocating their views. Additionally, most members are willing to support R&D on potential space weapons to a limited degree, especially if it means that money could potentially flow into their districts. Senators from states likely to benefit from missile defense–related money and jobs, such as Ted Stevens (R-Alaska) and Daniel Inouye (D-Hawaii), have been strong supporters. Whether the administration in office supports pushing the envelope of potential space weapons development, and the popularity of that administration, are also factors in how much Congress is willing to squint when scrutinizing the budget for potential space weapons programs.

During the Clinton administration, the benefit of the doubt regarding developing technology of ambiguous intent went against the military. The Clinton administration did not support the Clementine 2 program, an asteroid-intercept mission to collect data critical to construct weapons capable of deflecting asteroids or comets on a collision course with earth. The fear was that program supporters, including the Ballistic Missile Defense Organization, were simply using that mission as a Trojan horse to develop antisatellite capabilities, which the administration did not support. The Clementine 2 and ASAT technologies are basically the same: if one can hit an asteroid in space, one can hit a satellite. Those military space programs with weapons potential that did survive, and there were few, were usually, coincidentally, in states with strong congressional representation. Test facilities for a space-based laser (SBL) program were funded even during the Clinton years—built in Senator Trent Lott's (R-Miss.) home state.

Because space systems are very expensive, Congress is always very interested in them. Members of Congress able to bring government contracts to their districts will employ many voters, and employed voters are usually happy and grateful. Lott has maintained a close relationship with Lockheed Martin. As majority leader in the Senate, he supported the Theater High Altitude Area Defense (THAAD) program to the tune of $4 billion, despite its failing six of eight tests, before the program was killed in 2001.[45] Lockheed Martin pledged $1 million to the Trent Lott Leadership Institute at the University of Mississippi and made substantial campaign contributions to Lott.[46]

The relationship between Lockheed Martin and Lott is by no means unique. Nearly 60 percent of the seventy most important defense companies that won government contracts in postwar Iraq and Afghanistan had employees or board members who either served in or had close ties to the executive branch for both Republican and Democratic administrations, to members of Congress of both parties, or to the highest levels of the military.[47] As Republicans took control of Congress in 1995, adding $5 billion to $10 billion annually to the Pentagon budget beyond what President Clinton requested and then boosting it even more after the Bush administration took office, major defense contractors have favored Republican candidates on a two-to-one basis.[48] If winning a defense contract is good, winning a military space contract can be a trifecta.

As the Bush administration supports military space efforts, ambiguous military space projects are sustained unless Congress specifically rejects them. The Pentagon is vigorously pursuing small-satellite programs of the same variety as those being developed in China. China's satellites are a major cause of Pentagon heartburn for their technical potential as space weapons, though China purports to develop them for peaceful uses, including communication and remote sensing. But if China's small satellites can be space weapons, clearly U.S.-built small satellites can be also. There is a lot of effort at the margins of congressionally supported programs.

The military technology development and acquisition communities play their part in the fervent drive for technology development as well. Alfred Kaufman has traced the problem back to the ending of the Cold War. It was then, Kaufman states, that there was a collapse of the intellectual structure that had been erected to control the development of Western military technology: "That structure, which ensured that acquisition of military technology was guided by the specific operational requirements that flowed from a well articulated National Security Strategy, rested on the simple proposition that military technology should follow, rather than precede, the national security needs of the nation."[49] While national security strategies were not always articulate, and the need to tie technology development to a specific mission almost cost the world the GPS, Kaufman's general line of reasoning merits attention.

There has been nothing less than a quest in the military to be out in front, "leading change and challenging assumptions," as part of the DOD's transformation efforts. Network-centric warfare, defined primarily by the transformational connectivity of technology to benefit warfighters—those capabilities proclaimed on the lightning-bolt diagrams—has become the Holy Grail of military planning and acquisition. Flag officers are rewarded for supporting transformation efforts, and those who do not are soon gone. That has led to Pentagon adoption of business practices appropriate for staying ahead in a fast-moving market, but less appropriate for security policy with very real dangers of technology outpacing policy. In fact, according to Kaufman, "some technological innovations ... are beginning to follow their own self-generated developmental logic."[50] Add this phenomenon with the usually generous pots of money that come with fantastical ideas to a number of individual Air Force job descriptions that include looking for ways to expand space missions as part of the job responsibility, and it is no wonder both that elements in the Air Force and the DOD are pushing for space weapons and that industry is willing to oblige. The technology development programs that the United States can pursue to enhance force projection options are endless, as will be the tax dollars needed to bring them to fruition. Whether all are needed is another question. The U.S. military has historically been and remains fascinated with technology, often to the detriment of understanding morale and initiative. It has been suggested that "American security planners ought to consider ... less costly weapon systems rather than always planning for changes of greater technological complexity."[51] Unfortunately, just the opposite remains the norm.

Kaufman points out that one of the negative consequences of being on a perpetual quest for new technology is that it avoids dealing with the asymmetric state of the world. The technology-driven network centrists have redefined asymmetric warfare to mean "not warfare designed to counter our vulnerabilities, but rather warfare designed to rely exclusively on our own technological superiority. According to this definition, it is our asymmetric strength that really matters, not the enemy's asymmetric ability to avoid the effect of that strength."[52] This definitional slight-of-hand is essential when advocating for missile defense and space weapons, as countermeasures and responses to space weapons, which must be anticipated, are both easier and cheaper than either missile defense or space weapons themselves. Most important, the entire paradigm of technology development allows for the continued development of technology with little regard for either predictable responsive behavior from other countries or unintended consequences. As Christopher Van der Allen writes,

> Technology is just one part of the entire picture and though one side may have a superior weapon in one sector, countermeasures by the other side, which may be of an entirely nontechnical nature, may do much to negate any possible advantage. In this vein, American security planners might consider less costly countermeasures rather than match each technical challenge with one of greater complexity.[53]

Ideologically, the paradigm becomes techno-nationalism on steroids: building weapons systems to show that we can, hoping that they will be viewed as an indicator of power to subsequently deter all challengers.

Globally, military strategies increasingly rely on technology for advantage. In the United States, transformation is viewed as the key to lighter, more agile, and smaller forces that are able to overcome larger forces. But in all countries, military modernization is equated with technology. Space assets are a critical part of the technology that is, in turn, critical to military modernization.

The U.S. military is without a doubt the most powerful military force in the world, and space technology has successfully been a force multiplier for traditional forces. Its success in "see it" tasks has been substantial. More capabilities in that area—that is, toward monitoring both earth and space, including looking for debris, hazardous objects, and potentially nefarious activity—is clearly warranted. While capabilities in the "state it" areas have increased, the ability to provide information to the front lines still needs improvement. The "stop it" area is the most problematic, both technologically and politically.

The current U.S. approach to space assumes that any other country's efforts to use space for military modernization, or in some cases for economic development if dual-use technology is involved, is a threat to the United States. That assumption, and relying first and foremost on a technological response, is untenable in a globalized world. Other countries are seeking space technology for military and civilian purposes, and U.S. efforts to deny them will likely only result in increased determination on their parts.

As the sole superpower, the United States seems destined to shape the global order, and this can be done in several ways. Military force is one way, and the United States has pursued this option without hesitation. Relying solely on the tactics of hard power, however, is certain to trigger challenges. Increased opportunities for asymmetric challenges are presented as the United States increasingly relies on technology. Supporting international arms control regimes to balance the hard power approach has been rejected under the primacist approach to politics; it is being neglected to the peril of U.S. security. The United States is traveling a road to space dominance in a car without a steering wheel, and with no consideration of the terrain beyond immediate sight. The most dangerous areas where this is occurring are in space control and force application, the subjects of the next chapter.

The Weaponization of Space

FIVE

The best diplomat I know is a
fully activated phaser bank.
—Scotty, in *Star Trek*

When the Air Force announced in 1988 that it would regard space as a mission and not just a place, it reversed decades of tradition and doctrine.[1] Initially, space support and force enhancement were the missions associated with space, neither of them particularly controversial. Space would provide assistance to ground- and air-based warfighters. When space control and force application were added as missions in 1996, however, the Air Force began considering the potential for space as a battlespace arena, creating both the potential for expanding Air Force turf and capabilities and increasing funding to accomplish the latter and a plethora of other issues.

Force application is the overt weaponization of space, as compared with the de facto weaponization that has occurred under the guise of space control. Under either means, weaponization is very different from militarization. As discussed in the previous chapter, the militarization of space began even before civilian space programs did, as the military became interested in developing missiles, building surveillance satellites, and using satellites for military communications. But excepting the attempted foray into active weapons through the Strategic Defense Initiative (SDI), or "Star Wars," and its successor programs, the military use of space in the United States has been limited to passive systems. Weaponization crosses a threshold that all countries held inviolate for many years.

In July 2004, the British press began reporting fears that the United States was developing killer satellites capable of destroying European Galileo navigation satellites if it felt that potential adversaries could use them against the

United States.[2] The following October, another round of media reports surfaced, mostly in Europe, that the United States had threatened to blow Galileo out of the sky during a meeting at Whitehall on the topic of Europe's challenge to the American GPS.[3] It turned out that reporters completely exaggerated the issue, but the incident illustrates the distrust that other countries harbor about U.S. intentions in space.

How could the United States take out a satellite? As already pointed out, satellites are easier to hit than are bullets in flight. They fly in relatively fixed or predictable positions and are bright objects against a dark background, making them relatively easy targets. Extending the role of missile defense from purely defensive to power projection by, for example, using missile defense interceptors to preemptively take out a satellite meets the power-projection mission requirement to justify funding, as it sidesteps some of the program justification quandaries presented by the technical challenges of hitting a bullet with a bullet at 17,000 miles an hour. Other space technologies could expand U.S. options even further.

Those who support the weaponization of space meet challenges to their assertions regarding the need, viability, and effects of weaponization on U.S. and space security with shouting rather than substance. Supporters focus less on explaining the rationale for their horrendously expensive, technically questionable, and politically risky programs, and more on attacking those who question them as whiny, moaning, and wimpy fanatics. "It's as predictable as the rising and the setting of the sun," wrote the Heritage Foundation's Peter Brookes in June 2005, responding to reports that the Bush administration was considering overt support for space weapons in its new space policy. "Every time the United States moves to develop a new strategic weapons system that would improve national security, the left starts whining and moaning. . . . Arms-control fanatics are already condemning the new policy with frantic cries of 'arms race,' 'strategic instability' and militarizing space. . . . Fretting and fearmongering aside, the fact is that the 'final frontier' is critical to our national defense."[4] Nobody disagrees with the critical nature of space to national defense, but the issue demands serious analysis and debate, rather than sound bites and editorials.

Legal Parameters

Two treaties are relevant in discussions regarding space weapons: the Outer Space Treaty (OST) of 1967 and the now defunct Anti-Ballistic Missile Treaty

(ABM) of 1972.[5] The OST, ratified by ninety-six nations, including the United States, Russia, and China, has been referred to as both the Constitution of outer space and the Magna Carta of outer space. It provides a framework for legal considerations of outer space issues, but like many, if not most, international law documents, it is also laden with ambiguities. According to the OST, space is to be used for "peaceful purposes." That phrase contains the document's greatest ambiguity, and different treaty signatories almost immediately interpreted it to mean different things.

Traditionally, the U.S. view has been that "peaceful purposes" means nonaggressive, which is also open to interpretation. From the Kennedy administration until the Reagan administration, nonaggressive meant "passive systems only" and used for force enhancement. The Reagan administration, anxious to pursue active systems in conjunction with the SDI, defined "peaceful purposes" as defensive. While that ambiguous definition raised eyebrows internationally, it demonstrates the political influence on the legal frameworks. The U.S. military, not surprisingly, has always supported a more liberal interpretation of "peaceful purposes" as meaning nonaggressive and not necessarily restricted to passive systems. On the other side of the debate, a small minority of scholars have argued that the OST demilitarized space completely.[6]

Defining "peaceful" as "nonmilitary" creates problems as well. Japanese space activity was originally limited to "peaceful uses," with "peaceful" defined as nonmilitary. Over time, it became clear that, if the phrase were strictly interpreted, Japanese defense forces could not use satellites for even routine communications, weather forecasting, and reconnaissance. Such a narrow definition has also made it difficult, if not impossible, to embark on a joint endeavor deemed peaceful by the United States (allowing military involvement), but not by Japan. In 1986, during negotiations with potential partners in the International Space Station, the Pentagon suddenly decided that it wanted to maintain the option of the military conducting research on the space station. For the European Space Agency and Japan, ISS partners, that requirement was almost a deal breaker. Because both were legally prohibited from military space activity, including some fancy legal language in the memorandum of understanding between the partners was required to separate the Pentagon's potential role from those partners' involvement. Subsequently, countries that once interpreted "peaceful uses" narrowly have increasingly been reinterpreting it to expand their legal parameters.

The OST bans orbiting weapons of mass destruction (WMD), generally defined as nuclear, biological, and chemical weapons, but not weapons in general. Therefore, based on the OST alone, the door is open to legal differences of opinion on which types of space hardware are subject to international law and under what circumstances—for example, offensive versus defense purposes—and, more recently, whether preemptive attacks can be considered defensive. This issue is further complicated by Article III of the OST, which calls for space activities to accord with international law, including the Charter of the United Nations. This would suggest that military uses of space during wartime would have to be in self-defense, or authorized by the Security Council, to be legitimate.

To develop missile defense, the United States unilaterally withdrew from the ABM in December 2001. Before September 11, withdrawing from the treaty was a key political roadblock to missile defense; after it, withdrawal was accomplished with little fanfare or notice. The United States claimed that the ABM was no longer valid because the United States had signed it with the Soviet Union, and technically speaking, the Soviet Union no longer existed. Russian president Vladimir Putin, however, insisted that Russia, as the successor state to the Soviet Union, inherited its treaty obligations. Though negotiations with Putin to amend the ABM had been under way and were considered promising, after September 11, the United States no longer considered even the pretense of public diplomacy necessary, and simply withdrew unilaterally in December 2001.

Counterspace Operations in Context

The current vision for U.S. military space dominance was laid out in 1997 in the U.S. Space Command's *Vision for 2020* and has been more or less consistent ever since: among U.S. Space Command's goals are "dominating the space dimension of military operations to protect US interests and investment" and "integrating Space Forces into warfighting capabilities across the full spectrum of conflict."[7] What have been viewed as appropriate means to achieve that goal, however, has changed. With force application always a political lightning rod, weapons advocates have preferred space control as the route to space weapons. Space control involves efforts in four areas: surveillance, protection, prevention, and negation—with negation creating the most external concern. The Air Force has not wavered in pushing for broad interpretations of what is legitimately within the purview of space control. Consequently, it has been left to the administration in office to define the parameters. Assistant Secretary of the Air Force (Space) and Director of the National Reconnaissance Organization (NRO) Keith Hall, talked about the Department of Defense's 1999 space control technology program at a hearing before the Senate Committee on Armed Services on March 22, 1999, after an earlier classified briefing on the program. He said, "We are also planning to pursue negation technologies that could lead to capabilities that have localized, temporary and reversible effects as part of our broader information and force protection capabilities."[8] Even within the Air Force options in 1999, physically destroying satellites was not the preferred approach. The Clinton administration favored passive systems over negation within space control.[9] The doctrine promulgated in 2004 without dispute by the Air Force in *Counterspace Operations*, however, extends the scope of possibilities available within space control to include active systems. Thus the two previously separate space missions, space control and force application, begin to merge.

Surveillance refers to the ability of the United States to "see" in space, primarily through the Space Surveillance Network (SSN), operated through U.S. Army, Navy, and Air Force ground-based radars and optical sensors at twenty-five sites worldwide. Since the launch of *Sputnik* in 1957, the SSN has tracked more than 24,500 space objects orbiting the earth and is currently tracking more than 8,000 orbiting objects. Many others have reentered the earth's turbulent atmosphere and disintegrated, or survived reentry and hit the earth. Those objects regularly tracked, however, according to Strategic Command (STRATCOM), are not tracked continuously, but instead through "spot checks" because of the limited number of SSN sensors and their capabilities and geographic distribution. The space objects now orbiting the earth range from satellites weighing several tons to pieces of spent rocket bodies weighing only ten pounds. About 7 percent of the space objects are operational satellites; the rest are debris. The SSN tracks space objects 10 centimeters in diameter (baseball size) or larger. That means that there is some uncounted number of space objects that are not tracked or identified. The SSN identifies and catalogs space objects, warns NASA of potential collisions between objects and the space shuttle or the ISS, provides timely notification to U.S. forces of satellite flyovers, monitors space treaty obligations, and conducts scientific and technical intelligence gathering. The SSN would provide targeting and damage assessment information to support space control operations. It is a passive asset.

If anything has been learned in the post–September 11 world, it is the value of solid information. We need more and better information about what is going on in space. Currently, the SSN is the best available source of information for its task. Regardless of how "space control" is implemented, surveillance is a key part. If developing and deploying space weapons of any variety, fanciful or not, is even being considered, good surveillance must be given a high priority. Given the inherent vulnerabilities of space weapons because of the space environment, and the subsequent potential of a lose-or-use mentality in a crisis situation, the need for situational awareness cannot be overstated. Imagine the ramifications if the United States decided that a hostile action on one of its satellites was imminent and used a space weapon to destroy another country's satellite—only to find later that the "threat" had actually been space debris.

Crises are crises by virtue of quick decision making cycles and the potential for grave consequences. What if an American or a Russian early-warning satellite was hit by a piece of space debris, and the hit was interpreted as an attack? Would the nuclear response be considered?[10] As one Air Force official, acting as the enemy commander in a space war game stated: "[If] I don't know what's going on, I have no choice but to hit everything, using everything I have."[11] Situational awareness is considered an imperative in all military operations; whoever has the greater battlefield situational awareness is considered to have a strong military advantage. Yet improving the SSN has not received nearly the priority that other programs that depend on SSN capabilities do.

Regarding protection, the emphasis before the Bush administration was on the survivability and endurance of DOD systems, using passive methods such as ground station protection, satellite proliferation, hardening, communication cross-links, encryption, communications security protection, and threat warning sensors. The reasoning was that destruction would undercut U.S. commercial interests that depend on global cooperation, such as the international allocation of transmission frequencies required for effective communications. Previous policy recognized that purely unilateral U.S. action regarding working in space was impossible. Additionally, the high probability of collateral damage to U.S. and friendly space systems from debris discouraged kinetic kill programs. Destroying a satellite in orbit creates a lot of debris, each piece of which in effect becomes an indiscriminate small kinetic energy weapon, dangerous to its surroundings. The increasing use of commercial space assets for defense applications was also noted—a critical point, in that collateral damage to any commercial satellites could subsequently have a significant negative effect on the military.

The Clinton administration paid considerable attention to prevention as the newest aspect of space control. This included measures to preclude an enemy's ability to use data or services from U.S. and friendly space systems and services for hostile purposes. The technical answer to that problem was seen as encrypting satellite control and payload data to prevent unauthorized access. That included using licenses to monitor commercial satellites, ensuring encryption devices on, for example, remote-sensing systems. Those encryption devices were to create what is known as "shutter control," which, theoretically, would allow the U.S. government to prevent unauthorized access and limit data collection or distribution during periods when national security could be compromised by delivering images to hostile forces. Exercising shutter control, however, turned out to be problematic.

The difficulties with imposing shutter control became painfully obvious during Operation Enduring Freedom in Afghanistan. The Pentagon was concerned that 1-meter commercial imagery from the commercial *Ikonos* satellite would or could be obtained and used against U.S. troops in the same way that the United States employed such data against adversaries: for reconnaissance, targeting, battle assessment, and other functions. Rather than trying to exercise shutter control, however, the Pentagon chose to buy all the commercial imagery of Afghanistan itself, on an exclusive contract of $2 million a month that lasted just under a year. The reasoning behind the decision was that imposing shutter control could have resulted in the filing of First Amendment lawsuits by the press or even citizens' groups, requiring the Pentagon to fight a battle in court as well as in Afghanistan. The lesson: technology is not the answer to every problem.

Comparing views on space control before and during the Bush administration requires acknowledging that American attitudes have totally changed, not only about military space hardware, but about U.S. goals for its potential use. To a significant extent, the importance of military space must now be considered within a new U.S. view of the world and its place in it. Simply put, the view represented in the 2002 national security strategy (NSS) of the United States, and generally supported in the 2006 version as well, though with a slightly lighter touch, is that the world is a generally nasty place, but that the United States can fix it. This view was carried forward on the wave of widespread political reluctance to criticize actions taken in the name of the global war on terrorism after September 11, and a willingness to support large increases in defense spending. Under the Goldwater-Nichols Act of 1986, a new administration is required to produce an NSS by June 15 of the first year of taking office and "regularly" thereafter. The legislative intent is both to provide a statement of U.S. strategy and to act as a policy foundation on which the Pentagon can then build its own national military strategy (NMS) and its force structure. It is not uncommon, however, that an NSS is released behind schedule and goes relatively unnoticed outside a small circle of policy analysts. As a legislatively mandated document, the potential for its being just another yawn-inspiring periodic government report is inherently high.

The multiple NSSs released by the Clinton administration received little fanfare.[12] Like most preceding NSS documents, they were considered rather perfunctory, and perhaps appropriately so. Maintaining strategic ambiguity regarding its approach to world affairs has been a time-honored method of sustaining U.S. flexibility. But the NSS issued by the Bush administration in September 2002 was a different matter. It spelled out, in some detail, America's global post–Cold War ambitions.

The 2002 NSS clearly stated the intent of the United States to perpetuate its military supremacy and a willingness to use military force to reshape the international order, including preemptive strikes against those considered even potentially a threat to the United States. It codified the so-called Bush Doctrine, first publicized in a speech delivered on June 1, 2002, to the graduating class at West Point.[13] The salient elements of the doctrine were preemption, unilateralism, strength beyond challenge, and the notion of extending democracy, liberty, and security to all regions. Preemption has always been a U.S. foreign policy prerogative, but the Bush administration clarified the circumstances for use: whenever the United States or its allies felt threatened by terrorists or by rogue states potentially producing or procuring WMDs. While supporting "a distinctly American internationalism," the strategy also supported the right of the United States to act unilaterally if an acceptable multilateral course could not be found. "Strength beyond challenge" refers to maintaining U.S. forces so that the United States remains the sole military superpower and can actively promote the extension of democracy, liberty, and security around the world.

Supporters hailed the Bush strategy as "right on target with respect to the new circumstances confronting the United States and its allies."[14] Others, including analyst Andrew J. Bacevich, pointed out that the 2002 NSS implies that "the only path to peace and security is the path of action." That action, according to Bacevich, requires the United States to "charge down that path until we drop from exhaustion or fling ourselves off the precipice fashioned of our own arrogance."[15] Regardless of opinion on the validity or viability of the vision, the 2002 NSS significantly heightened the role of the military in conducting U.S. foreign policy.

In 2001, Secretary of Defense Donald Rumsfeld began talking about a "1-4-2-1" military strategy for the United States, a strategy eventually codified in *The National Military Strategy*.[16] American military leaders and forces have three priorities: to win the war on terrorism, to increase the powers of the four individual services to fight together, and to transform the nation's military forces. According to the strategy, in accomplishing these goals, the military will be better able to protect the American people, prevent conflict and surprise attacks, and prevail in war. The "1-4-2-1" aspect enters through the expectation of the military being able to defend the homeland, deter aggression in four key theaters, "swiftly defeat" two aggressors simultaneously, and have the power to occupy and effect regime change in one of those aggressor nations. With a smaller military, this strategy relies heavily on technological advantages.

Missile Defense

Officially, missile defense falls into the category of space control as a defensive system. However, if a country can technologically accomplish missile defense, it can also use that system for force application missions. The potential dangers of blurring the line between space control and force applications are further intensified by the possibility of using the technology in preemptive, force projection situations, force projection here meaning unleashing the military element of U.S. national power from the continental United States to another part of the world.

There are several different types of missile defense. The general categories used until recently were theater missile defense (TMD) and national missile defense (NMD). In essence, TMD involves intercepting missiles at relatively short ranges to protect troops and small areas, and TMD development efforts have been generally accepted for some time.[17] The United States and Russia designated trajectories considered acceptable for TMD programs even while the ABM was in place. Meanwhile, NMD efforts have been far more controversial. Within NMD, different technologies can be used, from ground-based kinetic or directed energy weapons, to air- and space-based weapons, usually depending on where in a launch trajectory a missile is to be destroyed. During the Clinton administration, the systems were clearly distinguished from each other, and more emphasis (read: funding) was placed on TMD programs. The Bush administration dropped the distinction between TMD and NMD, referring to all efforts simply as missile defense (MD), and reversing funding priorities from TMD to NMD programs.

A truly robust missile defense system would be multilayered, providing potential defense against missiles at several points in their flight—flight time being a critical MD consideration. It takes approximately twenty to thirty-five minutes for an intercontinental ballistic missile to travel from Russia or Asia to the United States. Boost-phase missile defense attempts to stop missiles immediately after launch, when early-warning satellites most easily detect the rocket engine's heat plumes. Destroying missiles during the boost phase also means that the potentially considerable debris created falls back to the launch site, rather than on the area being defended. Technically, the systems most capable of boost-phase intercept would be air- or space-based, since depending on the type of missile, the boost phase lasts only from about 170 to 240 *seconds*.[18] Mid-course intercept, on which the current program focuses, attempts to stop missiles in their flight paths. This is where developing the capability for a bullet to hit another bullet while traveling 17,000 miles an hour becomes relevant. Finally, as a last resort, there are systems that attempt to take out missiles in their descent phases, though that would likely result in a rain of destructive debris close to or on the intended target area.

The technology essential to developing a successful missile defense system is rocket science at its most difficult. It not only presents original science and engineering problems, but requires integrating multiple parts of the system—including missile launch detection, tracking, and plotting an interception course for a bullet to hit a bullet—all in a span of minutes. What could go wrong? But the technical issues involved are matched in complexity by the politics that have surrounded the program.

Legal debates during the Reagan administration over Star Wars technology confirm the role of politics in the history of missile defense development. Government lawyers had concluded that to test and deploy a national missile defense system, the United States would have to abandon the ABM. Unwilling and, at that time, politically unable to walk away from the treaty, the Reagan administration decided that the only other alternative was to enlist other lawyers to creatively reinterpret the problematic portions of the treaty. Original interpretations of the treaty had clearly stated that testing and deploying ballistic missile technology was banned. Reagan administration lawyers, however, argued that the treaty covered only those technologies that existed when the treaty was signed in 1972, and therefore excluded "exotic" technologies being considered as part of SDI, such as lasers and particle beams.[19] This new interpretation of the ABM cleared the way for initial development and potential testing. Further, while the ABM apparently prohibited using directed energy weapons for missile defense, the same technology was not prohibited when used in conjunction with developing, testing, or deploying ASATs. That provided another potential loophole for exploitation. Missile defense technology enables the possessor to destroy, with either directed or kinetic energy weapons, other missiles. As pointed out, however, the ability to hit a missile transfers, with some modification, into the ability to hit a satellite. The technologies for ASAT, planetary defense, and missile defense are virtually the same. In

1986, however, Congress banned ASAT testing to stop an arms race for what it felt was a weapon of questionable effectiveness. Congress assumed that the military would continue research in the ASAT field, and by extension missile defense, but could go only so far without testing. Congress's assumption was that no country would rely on an untested weapons system.

Between the 1986 ban on antiballistic missile testing and the George W. Bush administration, missile defense was not forgotten. The George H. W. Bush administration proposed a system called Global Protection Against Limited Strikes (GPALS), announced at a Pentagon news conference in February 1991. The official who presented the $32 billion plan was Stephen J. Hadley, then an assistant secretary of defense, future deputy national security adviser to George W. Bush. The secretary of defense at the time was Dick Cheney. The first Bush administration believed that the demise of the Soviet Union allowed the emphasis in missile defense to shift from protecting the United States against an attack by thousands of Soviet nuclear missiles to protecting the United States and its allies against perhaps several dozen missiles of any origin. An integrated, multilayered defense system using cutting edge technologies, GPALS had an estimated cost of $59 billion,[20] and Clinton eventually canceled it. But the ideology lived on, kept alive through think tanks and Congress, primarily the Senate.

With no public outcry for missile defense, continual congressional support appears more driven by ideology than politics. While there was sometimes bipartisan support for missile defense, it was usually as a limited option to more aggressive alternatives. The push to keep the missile defense effort alive was clearly from the Republican right as a moral imperative, the idea being that if the United States faces a deadly threat, then there is a moral obligation to defend against it, regardless of technical difficulty or cost. Supporters in think tanks were driven as well.

Former members of government and the military often populate think tanks, forming in effect a shadow government linked to lobbyists and corporate sponsors through crossovers among their boards of advisers. Those that supported missile defense included the original SDI think tank, High Frontier, as well as the Heritage Foundation, the American Enterprise Institute, the Hoover Institution, Empower America, and the Center for Security Policy (CSP). The external links among think tanks that support missile defense include strong ties to defense industries. The CSP receives about 25 percent of its annual revenue from corporate sponsors, including missile defense industry beneficiaries Boeing and Lockheed Martin; it is credited with persuading Newt Gingrich (R-Ga.) to include missile defense in the 1994 Contract with

America, while Donald Rumsfeld was on the board of Empower America and was a CSP adviser.

Congress has seen missile defense champions in both the House and the Senate; Senators Bob Smith (R-N.H.) and Jesse Helms (R-N.C.) were among the strongest advocates. In the House, inclusion in the Contract with America, a document introduced six weeks before the 1994 election, was considered a coup for missile defense supporters. Signed by all but two of the Republican members of the House, and by all the party's nonincumbent Republican candidates for that body, it laid out specific Republican plans for the future. The Contract with America was revolutionary in its commitment to specific actions. It was a triumph for Gingrich and the American conservative movement and provided momentum that carried through legislatively. For missile defense, that momentum translated into the passage of the Defend America Act of 1996, the National Missile Defense Act of 1997, the American Missile Protection Act of 1998, and the National Missile Defense Act of 1999. Congress pressed Clinton to proceed with missile defense, and when the White House did not respond, congressional advocates pushed it themselves. At times, both the House and Senate appropriation committees voted to allocate more funding for missile defense systems than either the president or the Pentagon had requested. At the same time, Helms was actively lobbying against renegotiating the ABM with Russia, favoring unilateral abrogation instead.

Threat assessments conducted before September 11 also provided a powerful impetus for missile defense. All missile threat assessments are based on basically the same numbers,[21] but sometimes reach vastly different conclusions. Is a glass half empty or half full?[22] Some analysts contend that arms control is an illusion and that things have gotten much worse.[23] Others contend that arms control has been successful, albeit slowly and incrementally.[24] The evidence supports both sides. As Albert Einstein said, "Not everything that can be counted counts, and not everything that counts can be counted"; as Rumsfeld is fond of saying, "There are unknowable unknowns."[25]

Several issues complicate missile threat assessments. The chances of an entire country being annihilated have likely decreased since the end of the Cold War; the chances of losing a city through a rogue missile attack or an accident have probably increased since the global war on terrorism began. Even the premise that the chances of country annihilation have decreased can be challenged, considering the India–Pakistan dispute, which now involves nuclear weapons on both sides. Are these assessments on which policies and actions of governments are based—usually from intelligence or military organizations—worst-case scenarios? Is that appropriate, and even expected for planning purposes, or does planning from a worst-case scenario trigger potentially escalating responses? More chilling, are the worst-case-scenario assessments clinically accurate? Could they be conservative?

Recent focus on whether or not organizations charged with making these assessments do so under political pressure raises a set of other problems. Was flawed intelligence on WMDs in Iraq a function of political pressure? George Tenet, former director of the Central Intelligence Agency (CIA), vehemently denied such charges, though some analysts had complained to the contrary. Skeptics wondered about the unusual nature of Vice President Dick Cheney's multiple visits to the agency before the war,[26] and the creation of an in-house DOD intelligence unit called the Office of Special Plans to circumvent the intelligence agencies. Intelligence reporting can be more or less neutral, but the analysis is inherently subjective because it is done by humans. Therefore, politicization is perhaps inherent as well, with degree becoming the only question. When policies and programs are on the line, "calling it as they see it" gets harder and harder for analysts to do.

Every year, the National Intelligence Council (NIC), whose members include the CIA and other security agencies, produces a national intelligence estimate (NIE). The 1995 NIE flatly stated that "no country, other than the major declared nuclear powers, will develop or otherwise acquire a ballistic missile in the next 15 years that could threaten the contiguous 48 states and Canada."[27] That estimate used parameters that were long-held standards. They included looking at threats to the forty-eight continental states; defining ICBMs as similar to those possessed by the United States and Russia—sophisticated, powerful, highly accurate, and ready to launch at a moment's notice; defining development of a long-range missile in terms of deployment; and looking at what was probable, rather than remotely possible, independent of significant political and economic changes.

Even before the report was officially released, those on both sides of the missile defense debate were acutely interested. Those against missile defense welcomed its findings; those for missile defense saw it as skewed to support Clinton administration policies. Curt Weldon (R-Pa.), chairman of the Subcommittee on Military Research and Development of the House Armed Services Committee and a staunch supporter of missile defense, took up the campaign to correct the skew.[28] Dissatisfied with the findings generally and feeling that they were politicized, Weldon called for further CIA briefs. Weldon was particularly upset that Alaska and Hawaii were excluded from the study. In technical terms, ICBMs had traditionally been considered capable of traveling distances of 6,000 miles. A North Korean missile would have to travel nearly

6,000 miles to hit California. The distance from North Korea to Alaska, however, is only 3,700 miles. That meant that missiles able to hit Alaska but not California were not considered ICBMs. Weldon felt that consequently, the missile threat to the United States was being underrepresented.

The political aspect concerned an American-Israeli-Russian issue. Israel was actively trying to get the Clinton administration to confront Russia about its missile experts working with Iran on a scaled-up Soviet Scud missile, known as the Shahab-3, which theoretically could hit Tel Aviv from launch pads in western Iran. The Clinton administration, anxious to cultivate relations with the fledgling Russian democracy and President Boris Yeltsin, was reluctant to do so. This reluctance to confront Russia, it was felt, at least partially motivated what some saw as minimizing the threat from new missiles.

The Republicans' initial attempt to get the CIA to amend its 1995 findings was unsuccessful. Then in 1996, former CIA director Robert Gates headed a blue-ribbon panel to look at the findings and report to Congress by December. That report was even stronger in its technical case against rogue states acquiring ICBMs in the foreseeable future. Undaunted, Congress appointed another commission, this one headed by Rumsfeld, called the Commission to Assess the Ballistic Missile Threat to the United States. Rumsfeld has chaired two commissions, the Ballistic Missile Threat commission in 1998 and the Commission to Assess United States National Security Space Management and Organization, or Space Commission, in 2001, and has been on record as supporting missile defense since the days of Ronald Reagan and Star Wars, and as a presidential candidate himself in 1987.

The Ballistic Missile Threat commission began its work by redefining what constituted an ICBM. Whereas previously, missiles with a flight range of 6,000 miles or greater—basically, the same distance as North Korea to California—were considered ICBMs, the new definition of an ICBM included virtually any rocket capable of landing any type of warhead anywhere on U.S. territory. Since the distance from Hawaii to California is more than 2,000 miles, that meant that missiles capable of traveling a much lesser distance than 6,000 miles, which had been counted as intermediate range ballistic missiles, were suddenly considered to be ICBMs. The new definition included as an ICBM the North Korean Taepo Dong-1 launcher,[29] which was tested once in August 1998 carrying little more than a radio transmitter and blew up after no more than 1,000 miles of flight time.[30] Under the new parameters, without any increase in hardware, the number of ICBMs worldwide immediately increased, simply because they were being counted differently. Threat assessments likewise suddenly showed a marked ICBM threat to the United States.[31]

Rumsfeld also rejected what he considered to be a major error in CIA estimates: mirror-imaging, which is the assumption that because it originally took the United States ten to twelve years to develop missile capability, the same time line would hold for countries developing the technology today. By contrast, the members of the commission spoke to engineers at U.S. defense contractors, including Lockheed Martin and Boeing, asking them how long it would take them to build an ICBM from the starting point of a Third World country. The answer they got was five years or less. Critics argue that Rumsfeld may have created a reverse mirror-imaging problem by assuming that Third World countries have the same access to industrial base requirements as do major American defense contractors. Not surprisingly, the unanimous findings of the Rumsfeld Commission were much more alarmist than the conclusions of either the 1995 NIE or the Gates committee. It predicted that a rogue state would be able to "inflict major destruction" on the United States "within five years" of its decision to develop an ICBM. Further, the United States might not even be aware that such a decision had been made for much of those five years.

Democrats had been allowed to appoint three members to the nine-person Rumsfeld Commission, so the unanimity of the report made the findings difficult to contest. Rumsfeld clearly understood the importance of a unanimous report and, some argue, narrowed the commission's mandate to ensure that outcome.[32] That is, the report assessed only threat, not the need for a missile defense system. Renowned physicist and commission member Dr. Richard Garwin later stated: "It's just that it's [the commission report] cited as the reason for building a missile defense, and I believe that was also the reason that the Rumsfeld Commission was created, because the advocates of missile defense in the Congress said, what we need is somebody who will endorse the threat against which we can build a missile defense and then our cause will be advanced."[33] In that regard, endorsing a missile threat to justify missile defense, the conclusions of the commission have been elevated to quasi-doctrinal status. It is rarely, if ever, noted that the technical parameters and definitional benchmarks used in the study were significantly lowered from those of the past to achieve the desired numerical results.[34]

By September 1999, there had been a mass conversion in the intelligence community. Politically pressured or not, "it was the largest turnaround in the history of the [intelligence] agency, and I was part of making it happen,"[35] Weldon stated proudly, clearly feeling that he had corrected past wrongs. The rules and parameters of threat assessments had completely changed. Beyond all the technical parameters changed by the Rumsfeld Commission, the CIA methodology for assessments soon went even further, abandoning the view that a lengthy testing period would be required before a new missile system could be considered a real threat. With the accumulated changes in methodology, the unclassified 1999 NIE section on ballistic missile threats concluded that over the next fifteen years, the United States would be most likely to face ICBM threats from Russia, China, North Korea, probably Iran, and possibly Iraq.[36] Based on these assessments and the political support they generated, by 1999, the United States started to pour billions of dollars into a missile defense system.

The initial post–September 11 popularity of the Bush administration generally, and Rumsfeld specifically, all but enshrined the results of the Ballistic Missile Threat commission and the new threat assessment methodologies. One might think that with all the pressing and competing needs concerning protecting the country after September 11, and the limited resources available to address those threats, the idea of spending billions of dollars on a questionable system based on questionable threat assessments against an unquestionably narrow threat might be reassessed. That has not been the case, and a reassessment is not planned. Missile defense has taken on a life of its own, as its supporters intended.

For its part, the military has had a love–hate, passive–aggressive relationship with missile defense for many years. While the military sees the potential for extra cash flowing its way for missile defense development, it wants to make sure that the money budgeted for development is new, not reallocated from more traditional service capabilities. Given the choice between developing a missile defense and building a new fighter jet, a new aircraft carrier, or a new tank, all the armed services would opt for the latter. But the potential for new money led to the creation of a Pentagon "Tiger Team" to formulate a variety of options for pursuing missile defense. The one that carried the day in 1996 was called the "3 + 3" plan: three years for development followed by three years for deployment. That was followed with including missile defense in the 1997 Quadrennial Defense Review (QRD). That report stated that national missile defense would need large infusions of funding to be ready for deployment by 2000, in accordance with the 3 + 3 plan. The "large infusion of cash" part got the attention of stressed aerospace industries.

Facing congressional pressure, President Clinton agreed to move forward with a national missile defense system based on the 3 + 3 plan, signed into law as the National Missile Defense Act of 1999. That action was viewed as a compromise after he vetoed several more expansive Republican proposals. The act legally required fielding an NMD system "when technologically feasible." Part of the pressure to acquiesce to congressional Republican pressure on missile defense was timing. The Monica Lewinsky scandal was breaking, impeachment proceedings followed, and the presidential election was not far away. Clinton did not want support or nonsupport of missile defense to be an election issue. The pressure was such that in January 1999, Clinton pledged $6.6 billion for NMD over five years.

Meanwhile, questions about the state of the technology and the schedule for deployment were increasingly being raised. The Commission on Reducing Risk in Ballistic Missile Defense Flight Test Programs, also known as the Welch Commission after its chair, General Larry Welch, the former Air Force chief of staff, was tasked to independently review testing practices for hit-to-kill missile defense interceptor programs, assess their adequacy, and identify best practices for the NMD program against specific program recommendations.[37] The first Welch Report in 1998 was followed by a second in November 1999. Both voiced strong concerns about the pace of the deployment scheduled, calling it a "rush to failure," and recommended delayed deployment. Consequently, Secretary of Defense William Cohen decided to delay deployment from 2003 until 2005, while stating in a third report published in July 2000 that a 2005 date would still be very risky.

The Clinton administration had agreed to develop and deploy an NMD system if tests could confirm its technical capability. The developers certainly tried. In tests in which the interceptor zeroed in on the brightest thing in the sky, the target was illuminated. The target also transmitted a signal for the interceptor to home in on. Even these carefully scripted tests did not go well. Of the three ground-based interceptor tests conducted between 1999 and 2000, only one could be counted as a success, and that success was more luck than skill. Initially, the kill vehicle had been unable to find the mock warhead and began to home in instead on a bright balloon decoy. As luck would have it, the balloon and the warhead were close enough that the warhead appeared in the field of vision of the kill vehicle, which then found it. The Pentagon's director of operational testing and evaluation subsequently stated that there was no basis to classify the test as either a success or a failure, because it was unclear whether the kill vehicle could have found the warhead were it not for the fortuitous position of the decoy balloon.[38]

There have also been allegations of test-data manipulation by the companies and organizations that benefit from the continuation of the program. Ted Postol, a Navy science adviser in the Reagan administration and a physics professor at the Massachusetts Institute of Technology (MIT), sent letters to Congress and the White House regarding a 1997 missile defense test and consequent study by Lincoln Laboratory, a federally financed MIT research institute, alleging a cover-up of serious problems with missile defense technology. Postol first became known as a missile defense critic after the 1991 Gulf War, when he asserted that, contrary to Pentagon claims, Patriot missiles had shot down few, if any, Iraqi Scud missiles. Postol was originally ridiculed for his contentions, but Secretary of Defense Cohen eventually admitted that the Patriot had not worked.[39] Postol's later allegations stemmed from the case of Nina Schwartz, a senior engineer at TRW, who accused her employer of faking missile test results on a prototype antimissile sensor intended to distinguish enemy warheads from decoys. If the sensor could not accomplish this task, the credibility of the entire system was questionable. But TRW denied the charges, and a 1998 report by federal investigators cleared the company of wrongdoing. The report was done under the direction of Lincoln Laboratory, which received in excess of $700 million in federal work in 2003. Eventually, the Government Accounting Office issued two reports in 2002 concerning the 1997 missile defense test.[40] While the reports stopped short of casting blame on any one party, they did find that the 1997 test had failed, rather than turning in an "excellent" performance, as the contractors had described.[41]

Nevertheless, under political pressure, Clinton agreed to proceed, but at a "technically feasible" speed. When the Clinton administration officially stated support in 2000 for building and deploying a missile defense system, the justification was to protect the United States. The phraseology of questions in opinion polls conducted in the late 1990s as evidence of public support for a high-cost missile defense program reflected that justification. A poll conducted by the Republican National Committee in 1998 asked this question: "Recent reports say the Chinese have 13 long range missiles targeted at the west coast of America. Knowing this, would you favor or oppose an effective National Missile Defense system capable of defending US territory against limited ballistic missile attack?" Seventy-six percent answered that they favored or strongly favored missile defense.[42] The Pew Research Center for the People and the Press prepared a survey in June 2001 that showed much the same. Respondents cited three primary reasons for supporting missile defense: protecting against accidental missile launches, protecting allies, and inadequate current defenses. However, in that same poll, when asked whether a missile or terrorist attack (the latter using WMDs) was the greater threat to the United States, 77 percent of respondents regarded terrorism as the greater threat, and only 10 percent chose missile attack.[43]

When the Bush administration entered office, everything changed. Paced development and deployment of missile defense based on successful testing was jettisoned, and deployment was ordered to take place by September 2004, not coincidentally just before the next presidential election. Missile defense was to showcase the administration's efforts to protect the United States. Test failures were no longer an option. Before a December 2001 test, the media reported, "the head of the U.S. missile defense program said a planned test Saturday night will be considered a success even if an interceptor and dummy warhead failed to smash into each other 144 miles above the South Pacific. 'This is not a pass-fail test,' said Lt. Gen. Ronald Kadish, director of the Pentagon's Ballistic Missile Defense Organization. 'Success would be if we learned a lot and gained confidence for the next step.'"[44] Carefully scripted, the kill vehicle did intercept the target in that test. By December 2002, there were eight ground-based interceptor tests: five "successes" (including the earlier questionable success) and three failures. After a test failure in December 2002 and renewed concerns from skeptics about technical feasibility and the exorbitant government funding being spent, the Pentagon simply "postponed" further testing. The next test was not held until December 2004, after the election. It failed, too, attributed to "an unknown anomaly." According to Lieutenant General Henry Obering, the new director of the Missile Defense Agency, a "glitch" caused the test failure.[45] After the December 2004 failure, Senator Jack Reed, (D-R.I.), a member of the Armed Services Committee, questioned whether money spent by the Pentagon on missile defense might be better spent elsewhere, especially given the mounting costs of operations in Iraq and Afghanistan. By contrast, a spokesman for program advocate Senator John Kyl (R-Ariz.) said that "one bum test" would not alter support for the program.[46] Each interceptor test costs between $80 million and $100 million.

Besides the ground-based interceptors experiencing difficulties in testing, there is also a sea-based portion to the mid-course interceptor part of the missile defense program. Interceptor missiles are placed aboard Navy ships with improved versions of a system called Aegis, which uses radar to detect hostile missiles and direct on-board weapons to intercept them. After three successful though orchestrated tests, the plan was to have a rudimentary version of Aegis ready for use in 2005, while work continued on the test-plagued ground-based portion of the layered system. A fourth test was held in June 2003. It failed. Two subsequent tests, in 2005 and 2006, both succeeded in hitting their targets. Aegis, basically a theater missile defense program, offers the United States its best missile defense option at present.

Beyond the significant technical challenges remaining, however, are the also substantial command-and-control issues. While missile defense interceptors are physically based in the United States, within the geographic purview of Northern Command, their mission is part of Strategic Command. Strategic Command is responsible for U.S. nuclear weapon assets, as well as space operations, information operations, integrated missile defense, global command and control, intelligence, surveillance and reconnaissance, global strike, and strategic deterrence. Missile defense would be used in conjunction with a threat from some region of the world, a region within the responsibility of one of the four geographic combatant commands outside North America. The military has been holding conferences for years on the command-and-control issues—who controls what aspects of missile defense—but clear answers are still lacking in many areas.

It is important to point out what a perfectly operating multilayered missile defense system could protect against and what it could not. Missile defense could potentially defend against limited ballistic missile attacks. It could not protect against cruise missile attacks, however, and over seventy-five countries have cruise missiles, including systems capable or potentially capable of being launched from ships off the U.S. coast. Cruise missiles are relatively inexpensive and difficult to detect in flight because their heat signature is very low compared with that of ballistic missiles. They also fly very low, escaping radar. Missile defense would also not protect against WMDs brought into the United States across borders in cars or containers on cargo ships. While the technology for container inspection has improved since September 11, the sheer quantity of containers entering the United States means that less than 5 percent of them are inspected.

The George W. Bush administration came into office committed to missile defense. The 2000 Republican platform stated that "the new Republican president will deploy a national missile defense for reasons of national security: but he will also do so because there is a moral imperative involved."[47] While noble and hence difficult to argue against, it is impossible to prevent a nuclear attack on the United States through missile defense technology, just as it is impossible to guarantee that no further terrorist attacks will take place.

The U.S. government has an obligation to weigh the multitude of threats to the American people—including missiles, terrorism, energy shortages, failed and failing states, crumbling domestic infrastructure, rising costs and limited availability of health care, education challenges, and balancing investments—to protect against threats that are most likely, where the consequences are most grave, and where investment in "insurance" is most likely to be effective. Where is the greatest risk? Some analysts argue not to do away with one type of insurance to buy another, but realistically, that is what people do every day. A family might want the maximum coverage available for health, fire, homeowner, tornado, hurricane, flood, car, and sink-hole insurance (if the house is in Florida), but it has to prioritize how much it can afford to spend on each. If all grave threats were to be addressed without regard to risk, the United States would certainly be heavily invested in a planetary defense program, protecting the earth from a certain catastrophic collision with a near-earth object (NEO). The question of a NEO impact is not if, but when. However, because a catastrophic impact may not occur for a hundred or even a thousand years, the government is not heavily investing in protecting against it, and rightly so.[48]

In July 2004, the first national missile defense program interceptor was installed at Fort Greely, Alaska. The plan was to install ten interceptors by the end of 2004, six in Alaska and four in California, but only six could initially be mustered due to delays in every aspect of the program. Two were eventually positioned in California in December 2004. Each interceptor consists of a three-stage booster rocket and a kinetic kill vehicle.

For missile defense to work, more than just interceptor hardware is needed. Two satellite systems, one in a high orbit and one lower, are needed to detect a missile launch, track the missile, and guide the interceptor to it. Originally, these were called the space-based intercept radar satellites, SBIRS-High and SBIRS-Low. Unfortunately, both lagged behind schedule, faced technical difficulties, and went way over budget. The cost of SBIRS-High jumped from $1.8 billion, with a first launch scheduled in 2002, to more than $8 billion, with launch in 2006 at the earliest. By 2002, the cost of SBIRS-Low, perhaps the most technically challenging satellite program associated with missile defense, was estimated at somewhere between $10 billion and $23 billion, resulting in serious challenges to funding, even from supporters in Congress. The program was put on hold and the satellites were stored, later to be restructured as the Space Tracking and Surveillance System (STSS) program to prove that missiles can be tracked after the boost phase, and data passed between spacecraft. Those satellites are to be launched in 2007. The flight demonstration will cost approximately $860 million, with additional funding needed later to achieve operational capability. Without these systems, the interceptors have to rely on existing systems, built in the 1960s and 1970s. Because of the positions of those systems, they are less than optimal for detecting a North Korean missile headed for Hawaii, even though North Korean missiles are a key rationale for the system.

If all the components parts are completed and deployed, they still have to be tested as an integrated system, and both hardware and software require examination. In the fifteen to twenty minutes that the system is in use, millions of lines of computer code have to run perfectly. It is not technically impossible, with enough time and enough money, but the current missile defense system has not even been fully tested as an integrated system. It basically it consists of unproved interceptors in the ground in Alaska and California. These interceptors provide no defensive capability to the military or the public, and the public has already spent between $80 billion and $100 billion to put them there, toward an unknown ultimate expenditure in support of a dubious outcome.[49]

In June 2005, an outside panel chartered by the Pentagon concluded that the rush to deploy a national system in 2004 resulted in shortfalls in quality controls and engineering procedures; in short, the Pentagon put schedule ahead of performance. Criticism of the program after the multiple test failures prompted the Missile Defense Agency (MDA) to request the outside review by the three experts: William Graham, a former head of NASA; Willie Nance, a retired two-star Army general who oversaw development of the missile defense system from 1998 to 2001; and William Ballhaus Jr., who heads the Aerospace Corporation. Though much of the report remains classified, the summary released focused on the rush to deployment rather than the fundamentals of the system itself, though it did describe the program as "one of the most complex military systems that has ever been deployed."[50] Clearly, defense contractors are making a lot of money, with little return to the U.S. public. At what point does one cut bait and use the money elsewhere?

The rationale from supporters has been that some capability is better than no capability, and that the system will be completed in increments. Further, knowing that it is less vulnerable to missile attack or nuclear blackmail, the United States could feel freer to intervene around the world in support of its national security interests. Not only would the United States have the most powerful military in the world as a sword, but missile defense could provide the corresponding shield. Whether that is comforting or frightening depends on one's perspective.

The Republicans did not raise missile defense as a campaign issue in 2004. Before September 11, having an "operational system" of some sort prior to the 2004 election was considered imperative, to demonstrate follow-through on commitments to protect the country. After September 11, however, whether American voters would view continued expenditure for missile defense as a priority for protection was apparently a question best not asked.

Missile defense has been touted as an essential protection against potential irrational actors in the post–Cold War era. Can the United States trust that Kim Jong Il, the North Korean leader whom the *Economist* once put on its cover under the headline "Greetings Earthlings,"[51] will not do something stupid with a nuclear weapon? That question, and the potential value of missile defense, was put to an initial test on July 4, 2006, when North Korea launched six missiles, including one Taepo Dong-2 long-range missile, which failed in its first stage. For a second time in as many tests, the North Korea missile met an inglorious end, earning it the nickname "Go Wrong-2."[52] Generally, the test proved primarily to be another example of North Korea's propensity for adolescent attention-seeking behavior rather than an exhibition of technical prowess.

The response from the United States was to declare missile defense operational, though what that meant was left unclear. Was there any significant technological difference between the "operational" system and the developmental system that was in place the day before it was declared operational? Luckily, Northern Command officials were apparently quickly able to determine that the missiles posed no threat to the United States or its territories, and so did not actually have to fire an interceptor. Missile defense critics argued that was a good thing, as the chances of a successful intercept were slim at best. President Bush himself said that he thought the United States had a "reasonable" chance of shooting down a North Korean long-range missile.[53] Clearly, had the United States fired an interceptor and missed, the United States rather than just North Korea would have been revealed as an emperor with no clothes. Nevertheless some lawmakers—according to one report, those "looking to burnish national security credentials"[54]—used the incident as an opportunity to seek even more funding for missile defense, potentially up to $9 billion more annually.[55]

Current political commitment to the program is astounding. Traditional oversight required for other Pentagon programs has been virtually eliminated. In January 2002, responsibility for program monitoring was turned over to a committee of military officers and civilian officials approved by the very group advocating the program, the Missile Defense Agency,[56] which was established by Rumsfeld in 2002 by renaming the former Ballistic Missile Defense Organization (BMDO) to underscore its priority on the defense agenda. The head of the MDA reports to a Pentagon senior executive board, and the agency includes the service secretaries and the Pentagon acquisition chief. All of them are White House political appointees.

The missile defense budget is twice that of the Bureau of Customs and Border Protection in the Department of Homeland Security and twice that of the Coast Guard.[57] According to a 2002 Congressional Budget Office study, the system could ultimately cost taxpayers upward of $238 billion by 2025.[58] Other external estimates have placed the cost of all missile defense programs to be between $100 billion and $1 trillion by 2030.[59] The Pentagon has declined to make an official estimate. Clearly, trade-offs are being made, as spending money on missile defense means that other programs, military or civilian, are being sacrificed.

In the post–September 11, post-ABM world, Bush administration officials state that international cooperation on missile defense "presents an opportunity to bolster coalition alliance solidarity."[60] Since the late 1990s, Japan has been the most enthusiastic partner with the United States on missile defense R&D, with its investment potentially reaching $2 billion for 2004 and 2005. Japanese security policy, including in areas regarding military space previously precluded, was strongly influenced by a North Korean Taepo Dong missile that unnervingly flew over the Sea of Japan in 1998. After the North Korean missile tests in July 2006, Japanese officials even uncharacteristically talked about a preemptive strike on North Korean missile bases. Besides Japan, Washington lists India, Germany, Great Britain, Denmark, Italy, and the Netherlands as partners at some level on missile defense, with other countries enlisted as well.

The partners, however, are not necessarily joining the United States because they feel that it is morally imperative to do so; likely, they are joining for more pragmatic reasons. Japan is concerned with North Korea. For others, improved diplomatic ties with Washington and any economic perks that might come with that are incentives. But even the economic perks are indirect. Missile defense does not create many jobs. Its bulging budget goes mostly to a limited number of high-tech labs and specialized firms in the United States, or in states where sites are located, so there are few jobs to be spread around to stimulate local economies. But when asked to participate, the political cost of not joining must also be considered.

In 2004, President Bush approached the Canadian government to become a missile defense partner. A debate immediately ensued in Canada, since a majority of neither politicians nor the public approved of missile defense. Many Canadians regard missile defense as "inevitably bound to take wars into space, and those on planet Earth to the ultimate Armageddon."[61] But they also recognize the need to protect the massive amount of trade conducted with the United States. Such trade is more important to Canada than it is to the United States, and it is potentially jeopardized if relations with Washington sour.

Ottawa had already decided in the summer of 2004 to help Washington implement the system, by expanding the mission of the North American Aerospace Defense Command. That expansion allowed the transmission of satellite and radar data about incoming missiles from Canada to the Northern Command. Nevertheless, Ottawa eventually decided in 2005 not to fully participate in missile defense. Choosing to limit participation was viewed as gutsy in Canada and as a setback for Bush in the United States.

With the election of a Tory government under Stephen Harper in 2006, however, the subject of Canadian participation in missile defense was reopened. Philip Coyle has been among those encouraging Canada to stick to its position of not participating. From 1994 through 2001, Coyle served as assistant secretary of defense and director, operational test and evaluation, in the Department of Defense. He was the longest-serving director in the twenty-year history of the office. In a February 2006 interview with the *Canadian Press*, he stated, "The concept of missile defence is quite seductive . . . [but] it's destabilizing, it's incredibly expensive, and it doesn't work."[62] A Canadian change of policy, if one is to occur, remains to be seen.

A third interceptor launch site (besides Alaska and California) in central Europe is also high on Washington's agenda. However, even central European countries that are comfortable with Bush administration policies have reservations. Many central European countries supported Bush in the Iraq war, but noted that opportunities for postwar reconstruction contracts at first did not materialize for them, and then became unwanted due to security concerns. They are mindful of attitudes toward the United States among members of the European Union and potential retribution against central Europe for getting too cozy with Washington. If Washington offers a sufficiently sweet economic deal, however, chances are that the United States can buy more missile defense friends.

The traditional arguments against missile defense for defensive purposes have been, first, that the system will not work. Persistent problems with the current system verify the technical issues. With enough money spent, a bullet can likely be made to hit a bullet, but even then, a second traditional argument must be considered as well. Which country—and countries are still the only global actors with ICBM launch capabilities—is going to send a missile with a big return address on it to the United States? If China is the threat, then the countermeasures it could develop and employ are cheaper and technically simpler than the missile defense system it is trying to defeat, negating the deterrent value. A potential countermeasure against boost-phase intercept would be to shorten the boost phase (a so-called fast burn), thereby decreasing the chances of a successful intercept. Countermeasures for mid-course intercept include using decoys and "bomblets"—submunitions packed in cluster bombs—to overwhelm the system. Is the United States building a missile defense system to protect itself from North Korea? Rather than waiting to be reactive, with a questionable system, many analysts suggest a more proactive approach. Pentagon favorite Thomas P. M. Barnett puts North Korea at the top of the U.S. "to do" list.[63] His approach does not require missile defense, but it does require cooperation from China. Beyond countries, terrorists seeking to send a WMD into the United States seem much more likely to take a cheaper, easier route, such as using cruise missiles (invulnerable to missile defense) or bringing the weapon into the United States through its inherently porous borders, due to geography and the open nature of U.S. society.

Taken together, the two arguments suggest that missile defense may be an unworkable solution to a nonexistent problem. Additionally, if the higher threat lies with other probable courses of attack, prudent risk management dictates utilizing resources elsewhere. There have also long been concerns that missile defense is ultimately motivated to benefit the military–industrial complex. By this cynical logic, as long as missile defense does not work, and as long as there is an ideological commitment to making it work regardless of cost, money continues to flow to the program in substantial amounts. Robert Jervis pointed out in 1978 that "if arms are positively valued because of pressures from a military-industrial complex, it will be especially hard for status-quo powers to cooperate."[64] Defense contractors are making billions developing a technology that so far has shown limited technical success at best.

The traditional arguments against missile defense have not been negated by either technological advancements or a change in the global environment. If anything, the events of September 11 may have made the missile defense imperative even less defendable than in the past—but not for those who support the notion of missile defense being used for more than its originally stated, purely defensive purposes. Force projection may be the rationale behind the extraordinary commitment to missile defense. It has also been suggested that missile defense, in conjunction with a strategy of nuclear primacy, may explain Washington's missile defense strategy: "The sort of missile defenses that the United States might plausibly deploy would be valuable primarily in an offensive context, not a defensive one—as an adjunct to a U.S. first-strike capability, not as a standalone shield."[65] In either case, if defense is no longer the rationale for missile defense, a full explanation of the changed rationale, and another public-opinion poll, might be interesting.

Schools of Thought on Space Weaponization

There are four basic schools of thought on space weaponization.[66] The first argues that due to U.S. reliance on space for both military and civilian capabilities, space dominance is essential, as it represents the "ultimate high ground." The second says that weaponization is simply inevitable, and therefore the United States would be remiss not to prepare. The third school asserts the importance of space, but seeks to maintain the status quo, limiting the militarization of space to passive systems, or at least defensive systems, through arms control agreements. Finally, there is the space sanctuary school, sometimes known as the space doves. This group would like to have space off-limits to the military.

Space is the ultimate high ground: that is heard a lot from military officials drawing analogies between cavalry soldiers holding a hill as the best position to both fight and surveil others from. Similarly, if air superiority is required to win wars, then space is the ultimate altitude from which to gain military advantage on earth. Certainly on the ground and in the air, controlling the high ground provides clear advantages. While using space as the high ground is advantageous in some instances for the hardware that can be placed there—three communication satellites providing near global coverage of earth, for example—the hostile and unique environment of space negates other aspects of the analogy. One of the reasons that missile defense is perhaps technically more suited as an ASAT weapon than as a defense against incoming missiles is that satellites are relatively fixed or predictable targets. Airplanes can perform evasive maneuvers very quickly. Ground troops can protect and conceal themselves at the top of a hill. Space hardware, however, is right out there in the openness of space, traveling in a determined and knowable path. Navy captain David Hardesty juxtaposed the notion of "high ground" with that of a "sitting duck."[67] So while space as the high ground does offer advantages, those advantages do not come without risks to be considered as well.

Second, there is the "Space Pearl Harbor" threat, which argues that as air, land, and sea have become battlegrounds, it is inevitable that space will, too, and leaders would be remiss not to prepare for that inevitability. As General Joseph Ashy of the Air Force, head of the U.S. Space Command, said, "The United States will . . . eventually fight from and into space."[68] The premise became the headline from the report of the Space Commission, chaired by Rumsfeld: "We know from history that every medium—air, land and sea—has seen conflict. Reality indicates that space will be no different. Given this virtual certainty, the U.S. must develop the means to both deter and to defend against hostile acts in and from space."[69] With the change of administrations, the premise became implicit policy.

Inevitability arguments usually begin with a statement about the lessons that history teaches. History lessons are valuable, but history is not one dimensional; it has many perspectives and dimensions. It does not serve the public well to cite history that is imperfectly understood or interpreted, or viewed from only one perspective, to justify ideological policy goals.

Besides, what in history has proved to be inevitable beyond death and taxes? The domino theory argued that if Vietnam were to go Communist, then the rest of Southeast Asia was certain also to go Communist. That premise was deemed inevitable enough that the United States went to war to try to prevent it. Nuclear proliferation was considered inevitable after the United States used nuclear weapons in Japan. More than sixty years later, countries are still declining to take the nuclear route as not in their interest. Some say the American Revolution was inevitable; but if it was, why was there not a Canadian Revolution as well? Is war inevitable? Clearly, inevitability is difficult to predict and therefore not likely a good basis for a major policy shift.

Space as a sanctuary seems a moot issue. It is unrealistic to expect countries to relinquish extensive military space programs that they already have. That some countries are currently redefining "the peaceful use of space" to mean "defense" or involving "passive" military systems illustrates a trend in exactly the opposite direction. Most important, declaring space a sanctuary would, quite simply, not be in the best interests of the United States.

The arms control school argues that while space weapons might offer the United States a short-term advantage, in the long term, they would actually weaken U.S. security, and therefore should be neither pursued nor encouraged. It is perhaps ironic that the arms control school is invoking the traditionally more "realist" approach, under which countries pursue paths in their national interest, often defined in terms of maintaining "power." The rationale is simple. First, the United States has more space assets in orbit than any other country and, consequently, relies on them more than any other country, for both military and civilian services and applications. However, because spacecraft are governed by astrodynamics, and are therefore highly predictable, space weapons offer many options for adversaries that wanted to attack them. A marble-size piece of debris in LEO would hit a satellite with about the same energy as a one-ton safe dropped from the top of a five-story building, because of the velocity at which it travels.[70] If timed correctly, a payload of sand launched on a relatively unsophisticated missile and released in orbit could considerably damage, if not destroy, U.S. spacecraft passing through it. Many countries have that capability, or other equally effective and often low-tech countermeasure capabilities. Because space-based weapons are so vulnerable, a hair-trigger mentality of use or lose the weapons could develop. Jervis wrote about this problem: "When weapons are highly vulnerable, they must be employed before they are attacked."[71] Ironically, if a space weapon were used in space, it would create a debris cloud most dangerous to other U.S. space assets. Consequently, the United States gains nothing by having space weapons and potentially loses the most by using them.

Other arms control concerns focus on space weapons potentially being used for space-to-earth combat: using a laser, for example, to take out a North Korean launch pad as part of a preemptive strike. Space-based weapons, and lasers specifically, are viewed as the only even potentially viable method for destroying missiles in their boost phase, which is what a real missile defense plan would look like. The line that is crossed here, however, harkens back to the motive behind the OST: ninety countries felt strongly enough about avoiding the psychological issues involved with WMDs constantly orbiting overhead that they ratified the treaty. Some people and countries feel just as strongly about the potential to be targeted by a laser orbiting in space.

In 2000, the United Nations General Assembly voted on a resolution called the Prevention of Outer Space Arms Race. It was adopted by a vote of 163 in favor to 0 against, with 3 abstentions: the Federated States of Micronesia, Israel, and the United States.[72] On December 8, 2003, 174 nations voted "yes" on a variation of the previous United Nations resolution, this one calling for negotiations toward preventing an arms race in space.[73] Similar resolutions have been passed consistently for almost a decade, with growing numbers of supporters each time. Only four countries abstained in 2003: the United States, Israel, Micronesia, and the Marshall Islands. In 2005, the United States and Israel decided to no longer abstain: both voted against the resolution, and were the only countries to do so.

There are other rationales for space weapons as well. There is the argument that space assets, as global utilities, must be protected and that the United States is the only country capable of doing so. While there has been no evidence that any country, organization, or person has yet targeted a space utility, such as the GPS, it could happen. The question then becomes one of determining the best way to protect American space assets, not necessarily which space weapon to build.

One of the more original rationales for space weaponization comes from Everett Dolman at the School of Advanced Airpower Studies at Maxwell Air Force Base in Alabama.[74] Dolman's primary motivation for supporting weaponization is clearly to maximize U.S. hard power, but he also suggests that the current international law and treaty regime, including the OST, have inhibited growth in the commercial space sector by denying sovereignty and derivation of wealth from space. Dolman advocates that the United States essentially seize control of space, act as a benign hegemon, and provide incentives to commercialize space more rapidly. Seizing control could lead to weaponization, but in this case only as a way for the United States to ensure the steady growth of commercial space and civilian space exploration.[75] The weak link in that rationale is the belief that other countries would currently consider the United States seizing control of space as benign. Further, would the United States be protecting space for commercialization by all countries, or only the United States? There might be some difficulty getting other countries to believe that opportunities would be equally open to all, given that concerns already exist that the United States wants to hold back the economic development of some countries to maintain its economic strength, and because of security concerns about countries such as China.

Space weapons are neither risk free nor without accompanying issues. If the United States intends to use space weapons to deny an adversary the benefits of space, specifically other countries using space as a force enhancer, then it immediately faces the quandary of having to identify dual-use technology that is even potentially militarily sensitive and being willing to destroy it. It is also flawed logic to develop space weapons under the assumption that an SDI-type arms race would prove too costly for other countries to keep up. Adversaries would not have to keep up, but only mount a much more technically plausible and less expensive asymmetric response. This returns to Alfred Kaufman's argument that focusing on U.S. superiority runs the risk of ignoring the asymmetric possibilities to counter those superior technologies, to the detriment of U.S. national security.[76] Would having the United States be the first to develop and deploy space weapons induce others not to develop and deploy them? Perhaps, but they would almost certainly concentrate on ways to minimize the value of American space weapons—which again, could push the United States into a position of use it or lose it. It also forces the United States to constantly try to stay ahead of the cheaper and easier countermeasures developed against its space weapons. That puts the United States on a quest for invincible technology that is either impossible or obscenely expensive to create.

Ironically, the countries most capable of developing space weapons, China and Russia, have been the strongest advocates for a space arms control agreement, which casts further doubt on the argument that space weaponization is inevitable. The United States has refused this approach, citing past attempts at regulating dual-use technology that proved unenforceable and unverifiable. Just as China cannot tell if missile defense is intended as a defensive system or an ASAT, the United States cannot tell if China's Beidou navigation satellite system is intended to coordinate traffic in Beijing or guide weapons to targets. That does not mean, however, that there are not other arms control options worth considering.

Space Weapons

In the 1970s, science-fiction authors Larry Niven and Jerry Pournelle wrote about bundles of metal rods being hurled to earth from satellites in space.[77] The Air Force's transformation plan, dated November 2003, discusses hypervelocity rod bundles that could strike ground targets anywhere in the world from space. What was frightening in fiction is even more chilling as a possible reality.

The military, particularly the Air Force, should not be faulted for doing its job. The factions in the Air Force pushing for space weapons are charged with exploring all possible ways of using and protecting U.S. space assets. In the context of a tightening budgets due to the expenses of rebuilding Iraq, internal disagreement concerning the need for space weapons versus the need for modernizing more traditional assets have tempered the pace of development. But it is fundamentally up to civilian leadership to decide the future course of direction for the United States. Will the United States continue to develop space weapons through programs that evade congressional scrutiny by hiding in gray areas of ambiguous dual-use technology, or will the administration require a reasoned decision rather than allowing a de facto one? As long as nobody stops it, the military will proceed.

A 2004 report by the Federation of American Scientists (FAS) considered a wide range of U.S. space-asset vulnerabilities, concluding that space weapons are not preferred or even desired to address threats.[78] But Leonard Weiss, the chairman of the panel that wrote the report, stated at a press conference that the amount of R&D money being spent for space weapon–related programs created an "unstoppable momentum" toward deployment.[79] Although the dual-use nature of space technology, and the ambiguous intent often surrounding its development, requires that the United States be prepared to respond if other countries deploy space weapons, the R&D currently being pursued in the United States clearly extends beyond that level, and it is growing.

Technology wish lists are routinely incorporated into documents such as the Air Force's strategic master plans and transformation flight plans as posturing for funding increases. Thankfully, as Dwayne A. Day writes, the manned spacecraft propelled into orbit by nuclear bombs and the Orion space battleship, both once requested, were never approved.[80] But other programs are approved—even ones that, as Day also points out, perhaps do not "abide by the laws of physics" or "the laws of fiscal reality." Unless the brakes are applied very quickly, it appears almost certain that the United States will deploy space weapons.

Considerable time has been spent examining when the space weaponry line is crossed.[81] The Russians once considered the shuttle to be a potential ASAT. People in many countries outside the United States consider missile defense as an offensive system. I have heard it argued, not very effectively, that space-to-space weapons are space control, while only space-to-ground weapons constitute space weapons. The definitional bickering expands the gray area for allowable program development.[82]

At least some counterspace programs, intended for use against the presumably hostile space assets or activities of the enemy, are already found in the DOD budget.[83] In fiscal year 2004, the Pentagon asked for more than $82 million to develop counterspace technologies, and a total of over $352 million between fiscal years 2004 and 2009.[84] This money is in addition to the counterspace operations money that can also be found in the Air Force budget, sometimes within the space control items. The fiscal year 2005 budget includes funding for active research programs, including the hypersonic Common Aerospace Vehicle (discussed in chapter 4), which is potentially useful to launch space weapons into orbit on very short notice. Money for dual-use technology potentially amenable to space weapons is also sometimes tucked into the Department of Energy budget as well.

The R&D efforts most amenable to space weapons are connected to developing space-based antiballistic missile systems and microsatellites, or microsats. These programs can be defensive or, in the case of microsatellites, can even be used for nonmilitary purposes, and subsequently draw less congressional scrutiny. But realistically, given President Bush's December 2002 statement that the United States would continue the "development and testing of space-based defenses, specifically space-based kinetic energy interceptors (hit-to-kill) and advanced target tracking satellites,"[85] it takes nothing more than a couple of word changes to make those defensive hit-to-kill interceptors into offensive hit-to-kill weapons.

In the federal budget, the ballistic missile defense system (BMDS) interceptor program, a boost-phase kinetic energy interceptor potentially based on land, at sea, and in space, bears watching. While Congress slashed the budget request for the BMDS by $182 million in fiscal year 2004 and ordered the MDA to focus on land and sea basing, some programs seem to refuse to go away. The

Near Field Infra Red Experiment (NFIRE), a satellite to collect data on missile plumes, retains support to conduct two missile "fly-bys" to allow it a close look at a burning ICBM, though the House tried to terminate the program. In October 2006, Representative Terry Everett (R-Ala.) expressed disappointment that the testing of a missile-seeking projectile was dropped from the NFIRE program and advocated building another NFIRE satellite, to include a kill vehicle.[86] The Space-Based Interceptor Test Bed program to develop materials for a space-based interceptor is not dead either. There are also funds in the ballistic missile defense technology program that could be used to evolve space weapons capabilities, under the umbrella of the Advanced Technology Development (ATD) project. The remains of the space-based laser program canceled in 2002 at the behest of Congress now lives in the ATD project, awaiting revival or refocus under a new name.[87] Some of the most interesting programs within that project involve microsatellites.

Weighing less than 100 kilograms, microsats are widely attractive to many countries and organizations because they cost less to build and launch than do other satellites. They can be used for scientific experiments and applications. Universities have long had an interest in small satellites (smallsats) and microsats. Smallsats weigh under 500 kilograms; microsats are a subset of small satellites. The Surrey Space Centre in England, associated with the University of Surrey, has become a leader in developing and selling smallsats for countries seeking a relatively inexpensive entrance into space.

The potential benefits of microsat use, especially to militaries, extend beyond cost considerations. In a 1999 study on microsatellite technology requirements, the Air Force concluded that microsat development should be pursued as quickly as possible. Certain microsats can maneuver around and image other satellites—potentially even clandestinely—and even damage or destroy other satellites by smashing into them or using other techniques. The study called for satellites to intercept, image, and, if necessary, "take action against" a target satellite, meaning damage or destroy it.[88] The Air Force has three major microsat programs—MightSat, TechSat21, and the XSS Experimental Satellite series—with $68 million in the 2004 budget for them and $502 million over the next five years. *XSS-10*, the first of the series, was launched in January 2003. *XSS-11* followed in April 2005 to navigate larger distances and perform more difficult maneuvers. There is even discussion about potentially combining kinetic kill antisatellite technology with the XSS program, thus creating "killer satellites." Besides the Air Force, both the Defense Advanced Research Agency—where the most highly classified programs are developed—and NASA have funded microsat programs, potentially for such close-up operations as satellite refueling. Again, however, the ability to maneuver close to a satellite for a service task equates to the ability to maneuver close to a satellite for nefarious purposes. An Air Force official, speaking to a reporter from the trade publication *Space News*, stated, "XSS-11 can be used as an ASAT weapon."[89]

On the more exotic end of the spectrum is the "Rods from God" program, in which cylinders of tungsten, titanium, or uranium are hurled from the edge of space to destroy targets on earth, striking at speeds of about 7,200 miles an hour with the force of a small nuclear weapon—frightening, but amazing if it worked. The entire notion, however, of what is and is not possible in the physical environment of space has been largely neglected. The physical laws and technical confines of space weapons are largely not understood outside a segment of the science and engineering communities, and considered almost an annoyance by policy community advocates. David Wright, Laura Grego, and Lisbeth Gronlund provide scientific and engineering information necessary to make sound, informed policy.[90] If ideology supersedes facts, however, the question is whether anyone will care.

Beyond questions about the technical viability of the program, there are very real questions about why the United States would need it. What could the United States do with this capability that it could not do faster and cheaper with a ballistic missile? Proponents suggest that such a weapon could be used to destroy a ground-based laser site before it could damage a U.S. satellite, but physics dictates that even a kinetic energy weapon such as an orbiting long-rod projectile would take some ten minutes to arrive at a suborbital position, and five minutes to fall from an altitude of 450 kilometers.[91] By that time, the ground-based laser could have done its damage—unless the space-based weapon was used preemptively. The primary attraction of such a weapon appears to be its ability to provide the force of a nuclear weapon without the radiation. Pursuing these programs apparently is intended to perpetuate an image of the United States as untouchable. It might do that. It definitely, however, also perpetuates an image of the United States as a space-age aggressor, putting it at odds with the rest of the world in yet another area.

In "persuading" countries to work on missile defense, raising the possibility of shooting down a European navigation satellite, and developing space weapons technology, the United States presents an impression of unhesitating willingness to use hard power against any perceived obstacle to space dominance, however nebulously defined and whether explicitly or implicitly stated. In a role reversal from historical precedent, "cooperation" with the United States on missile defense neither generates soft power nor eases concerns about U.S.

ambitions in space. In some cases, quite the opposite is likely true. Whether achieving and maintaining space dominance precludes other countries' right to use space, even for military purposes, remains an unanswered question. Perhaps an even more important question is whether such a definition of space dominance is in the best interest of the United States. The assumption now is unquestionably that it is.

Space technology offers both opportunities and the dilemma of dependence on that technology. The doctrine outlined in *Counterspace Operations* sets the United States on a course in which it demands control over all space hardware and everyone else must have faith in its benevolent intent and reasoned actions. Whether that is likely, or whether it will trigger unintended consequences detrimental to U.S. national security, should be considered carefully. So far, that has not happened.

The United States must protect its space assets, but relying solely on technology and hard power is not the best way to accomplish that goal. It is imperative that the U.S. military remain technologically superior, but it is not economically viable to pursue every program alternative, though it is clearly profitable for defense contractors. In deciding which technology programs to pursue, choices should be based on a comparison of threats, need, and technical feasibility, not on ideology. Exclusive reliance on hard power will unquestionably result in other countries feeling threatened, and the consequent development of asymmetric challenges to our technological "fixes" must be expected.

Finally, it is unclear whether all the expensive technology being developed is the best or even a viable way to meet the goals for which it is intended. If the goal of missile defense is to protect the United States from WMDs, then there may be a significant mismatch. Questions still remain regarding the need and technical viability for the system. A risk-based threat assessment seems warranted. But missile defense and space weapons have become bright shiny objects that the Bush administration seems fixated on, regardless of cost or potential effectiveness. Consequently, effects-based and rational analysis often becomes subjugated to political goals, with negative unintended consequences. The communication satellite industry is well acquainted with that situation, having been an early victim of the politicization of space.

SIX

The Politicization of the U.S. Aerospace Industry

We are jeopardizing America's
dominance of the satellite industry.
—Dana Rohrabacher

The space technology on which the U.S. military has become so dependent, and is so enamored with, is built by an increasingly small number of American aerospace companies. These companies have long led the global aerospace-industry sector in both technical know-how and sales. A healthy aerospace sector is a requirement for all the missions that the military wants to accomplish in space. Yet at least one part of the sector, communication satellites, has been the victim of the fallout generated from a government investigation in the late 1990s, largely influenced if not driven by politics—to the detriment of U.S. national security.

The multibillion-dollar U.S. aerospace industry leads the world in capability and had a trade surplus of $11.5 billion in 1998. Since 1990, it has been navigating an export-licensing procedure much like Alice navigates the upside-down world she encounters down the rabbit hole. This industrial sector, critical not just economically, but for the capabilities it provides to both the civilian and military sectors, has been the victim of partisan and bureaucratic politics.

The aerospace industry generally includes civilian aircraft, military aircraft, missiles, space, and related products. Space made up 24 percent of the total industry sales in 2004, with military aircraft composing the largest share. According to the Aerospace Industries Association, the largest aerospace industry organization, the U.S. aerospace industry generated $161 billion in sales in 2004, up 8 percent from 2003, with most of the increase attributable to robust Pentagon spending.[1]

Within the aerospace industry's space sector, there are further components. According to a study by the Futron Corporation, within the aerospace sector in 2001, telecommunication and remote-sensing services generated $1 trillion; satellite services, $6 billion; satellite manufacturing, $14 billion; ground systems, $20 billion; and launch services and vehicle manufacturing, $5 billion.[2] Market leverage is gained through inclusion in the food chain: satellite manufacturers often subsequently become involved with satellite services or ground systems.

Clearly, there is money to be made in the space sector of the aerospace industry, but the potential is not what it used to be. Aerospace used to be among the top three sectors positively contributing to the U.S. balance of trade. That trend began to take a dramatic downturn in 1998. The U.S. percentage of the world launch vehicle market dove from about 50 percent in 2000 to about 20 percent in 2002. The U.S. percentage of the world satellite market dropped from about 65 percent between 1997 and 1999 to 40 percent in 2001 and 2002. The lower the numbers get, the less there is to be made by American companies through food-chain involvement in the space sector. Aerospace employment figures tell a story of downturn as well. In 1990, over 1 million people were employed in the aerospace field. By 2004, that number had dropped by more than half.

While many sectors of the economy have been affected by changing markets and globalizing industries, the reasons for and effects of the change are unique in aerospace and carry significant security implications. In other sectors, jobs, the potential loss of critical skill sets, and economic growth are the primary issues. In aerospace, all these factors, plus the loss of capabilities considered essential to military operations, are at stake. Ironically, many of the negative effects felt by the aerospace industry since the mid-1990s have been the predictable but unintended consequences of government actions taken in the name of national security. In 1998, the United States dramatically changed its export-control laws on communication satellites and other aerospace technology. The official reason was to stop the spread of dual-use technology considered of military value. But the capabilities for producing that technology had already been globalized, largely due to U.S. efforts starting in the 1960s. The real reason was largely political.

Attempts to deny aerospace technology to other countries, for economic or political reasons, do not have a particularly successful track record. The United States lost its monopoly in commercial launch services to the European Ariane launcher after the United States tried to restrict the types of payloads it would launch for Europe, as discussed in chapter 2. Ariane captured

50 percent of the commercial world launch market by 2001. History has repeatedly shown that countries cannot be denied technology if they deem it essential to their futures. The speed at which the technology is acquired, the level of sophistication of the technology, and the suppliers countries rely on for services and parts might be managed, but given sufficient political will and enough money, the United States cannot stop other countries from developing or obtaining technology. As mentioned in previous chapters, the United States does not have a monopoly on science, physics, and engineering knowledge—just a big lead. It should have learned this lesson with the Ariane case, but it did not.

Between 1962 and 1998, the international communication satellite industry evolved, benefiting from nearly forty years of support from the U.S. government. The American founders of this industry foresaw many benefits for the world, and during this period many of these benefits became reality. Communication satellites provided global linkages for voice, data, and media transmissions. In 1980, the Cable News Network (CNN) became the first twenty-four-hour news channel, and by the 1990s, it was established as an almost instantaneous window to the world, made possible by satellites. But in the spring of 1998, after the *New York Times* published lurid articles alleging the illegal sale of American technology to China, U.S. domestic politics intervened. A congressional inquiry was formed with strong partisan overtones: the House Select Committee on U.S. National Security and Military/Commercial Concerns with the People's Republic of China, dubbed the Cox Committee after its chairman, Christopher Cox (R-Calif.).[3] The Cox Committee was charged with looking into two problems: allegations of technology theft at national laboratories, resulting in the arrest of Wen Ho Lee, and suspicions that American technology had been illegally transferred to China after a 1996 satellite launch accident in China. The Cox Committee's focus on whether or not Chinese rockets used to launch Western communication satellites could lead to improved Chinese offensive ICBMs changed the existing satellite export paradigm. Seemingly overnight, long-standing, good-faith efforts to bridge cultural and communication gaps with satellite communication were described as threats to U.S. national security. Communication satellites, as dual-use technology, became subject to the same government controls as military satellites, tanks, or guns, for purposes of sale overseas.

Lee, a scientist at Los Alamos National Laboratory, was arrested in December 1999 and held without bail in solitary confinement, charged with violations of the Atomic Energy Act. On August 24, 2000, Federal District Court judge James A. Parker released Lee after he concluded that federal officials had lied at the initial bail hearing, when they described the scientist as a threat to U.S. national survival. Parker issued an extraordinary public apology and declared that Lee's jailing "embarrassed our entire nation and each of us who is a citizen of it." Within weeks, its case in ruins, the government dropped fifty-eight of its fifty-nine charges against Lee. He pleaded guilty to a single count of mishandling nuclear information and was sentenced to time served. Noted Asia expert Chalmers Johnson compared the activities of the Cox Committee with that of an earlier congressional committee, one chaired by Joseph McCarthy.[4]

Investigation into the second problem—alleged illegal technology transfer to improve Chinese missile technology—was equally melodramatic, including sensational charges, sensational findings (largely refuted later), media hysteria, and fearmongering. As has already been stated, the entire premise of using rocket technology to improve missile technology is erroneous. Rockets must be precise due to the nature of their task—launching million- or billion-dollar satellites into orbit. Missiles need to exist as only a deterrent, and, like horseshoes, they do not have to be completely accurate to be effective. Further, rockets tend to fail not because the individuals launching them do not understand what they are doing, but because of very routine human error. That principle deserves a closer look to appreciate the technically challenged and politically trumped-up nature of the entire Cox Committee effort.

Rocket Science versus Rocket Engineering: Why Rockets Fail

Rocket science is the physics behind the operation of a missile or launch vehicle. The knowledge was developed over sixty years ago, and today, the basics of rocket science are contained in textbooks published in every conceivable language. Rocket science is not an American invention; it is almost anything but. It cannot be regulated by the United States without burning books worldwide.

Rocket engineering, however, is an art that was simultaneously perfected by the United States and the Soviet Union during the Cold War. It is very mundane and can be learned only through experience; the idea that it can simply be transferred from a technologically superior culture to an inferior one is based on a lack of understanding of what is involved at best, and smug superiority at worst. Rocket engineering is no more than incredible attention to detail. Possibly unique among all things created by humans, a rocket must work the first time it is used. No one would buy an automobile or a television unless it had been tested; yet if an automobile or a television fails to work, it can be repaired. Satellites are similar to launch vehicles in that they must operate in a harsh environment for ten or fifteen years without the possibility of repair. They are more forgiving, however, in the sense that if a component fails to work, the satellite will normally not fail catastrophically. Sophisticated redundancy schemes have been developed to replace any failed component and to allow transfer of operations to that redundant unit safely.

Consider some of the factors that caused the failure of satellite launches from 1990 to 2000:[5]

February 22, 1990. In the early-morning hours, representatives of two Japanese satellite service companies, the Space Communications Corporation and the Broadcast Satellite Corporation, waited nervously with representatives of GE Astro and Ford Aerospace, the manufacturers of their satellites, and flight controllers from Arianespace, in the Jupiter Control Center near Kourou, French Guiana. The launch site, carved out of the jungle in what was once a French penal colony in the northeastern corner of South America, was chosen because it is only 7 degrees north of the equator with a clear view of the Caribbean to the east, making it an ideal location for launching GEO satellites. Ten minutes after liftoff, the launch vehicle, with its cargo of two satellites, crashed into the sea, and shortly afterward, VIPs from the outdoor viewing stand ran in, wondering aloud about the toxic fumes wafting overhead.

The next morning, pieces of the satellites and launch vehicle washed up on the beach, and the recovery activities began. The cause of the launch failure was found a few days later—a cleaning rag left in a fuel line.

March 14, 1990. Intelsat and Martin Marietta controllers sat side by side awaiting one of the few commercial launches ever attempted by the Titan 34D, which normally served only the U.S. military. Commercial marketing of the Titan would prove to be short-lived, after the embarrassing failure of this *Intelsat 603* mission.[6] The launch appeared to be picture-perfect, but when the launch vehicle had done its job and the satellite controllers at Intelsat's Washington control center were to take over, it became apparent that something was wrong. Quick thinking on the part of the Intelsat launch personnel quickly diagnosed the problem—the upper stage of the Titan had not separated from the satellite, and if something were not done soon, the combination of the Titan second stage and *Intelsat 603* would burn up in the atmosphere. The launch vehicle had been designed to launch either one or two satellites, so it was equipped with two circuits designed to send electrical signals to separate the launch vehicle from the satellites. As a result of a communications failure between the hardware and software groups at Martin Marietta, *Intelsat 603*, which was the only passenger on this launch, had been connected to one circuit while the software was set to fire the other.[7]

After some involved technical maneuvers, the Intelsat controllers were able to put the satellite into a safe orbit, where it survived for over two years. Then, on May 7, 1992, the astronauts on the space shuttle *Endeavor* replaced a key component, enabling *Intelsat 603* to be successfully inserted into GEO. Although a remarkable engineering feat, the cost of the mission was comparable to what it would have cost to build and launch a replacement satellite. The *Intelsat 603* flight was only the Titan's second commercial flight; the failure, which was directly attributed to poor systems engineering and a failure to properly test the launch vehicle, triggered an ugly lawsuit for negligence on the part of Martin Marietta, the Titan manufacturer, and Martin Marietta soon exited the commercial launch vehicle business.

February 14, 1996. Intelsat 708 was launched from Xichang, China, leading to the Cox Committee and new export-control regulations. The Chinese Long March rocket lifted only a few feet off the ground before tipping over and scoring a direct hit on the hotel that housed all the customer personnel. The Chinese government eventually acknowledged six deaths, though Westerners who were present estimated the death toll as many times that.

After months of effort, Chinese engineers were able to simulate a gyroscope component failure that would result in the exact symptoms that had been observed in the failed launch. This led them to inspect a gyroscope in their storeroom; they found that one of the solder joints was improperly designed and suffered from a well-known malady called the Kirkendall effect, a metallurgical defect that has been recognized by engineers since 1947. Had that unit been flown on the next launch, it too would have failed. There was speculation in the report of the Cox Committee that the Chinese would not have found the cause of the failure unless they had been given incentive to do so by the Loral Independent Review Committee, which had been established to provide assurances to Loral Space and Communications, Long March's customer, and Philippine Long Distance Telephone Company, the next Long March customer, that the proper corrective action had been taken.

In the end, even the Cox Committee concluded that Loral had not transferred technology to China, but that it had provided a defense service simply by being skeptical of the initial incomplete explanation of the failure. The ultimate irony of the *Intelsat 708* flight is that Chinese engineers already knew about the source of the problem. The fix was easy because they had made the same fix to five other such solder joints; somehow, they had missed the same problem in this sixth location.

May 5, 1999. The first two Boeing Delta 3 flights ended in failure. The second Delta 3 flight, with the Loral *Orion 3* satellite on board, failed even after exhaustive internal and external review boards had examined the first flight failure in 1998. This time, the cause was the Pratt & Whitney (U.S.) RL-10 engine, which powers the second stage of the Delta 3. Using the RL-10 engine, a workhorse since 1962, seemed to be a very low-risk decision for the upper stage of the new Delta 3 as well. However, the engine had undergone hundreds of engineering design changes during its lifetime—all "process improvements"—to, for instance, make it easier to build or to meet new environmental safety regulations. Although there is still debate within the engineering community, the straw that broke the camel's back after all these changes may have been minor dimensional changes to a nozzle. This led to the need to manufacture a new tool to hold the pieces of the nozzle together while the pieces were brazed together.[8] Subtle changes in the dimensions of the nozzle resulted in an incomplete braze joint over the entire length of the nozzle. Couple this with a misinterpretation by production personnel of the braze inspection criteria, which themselves had been written years before by someone in engineering, and the Loral *Orion 3* spacecraft ended up in a useless orbit.

In 1998 and 1999, there were so many embarrassing and expensive U.S. launch failures that President Clinton ordered a blue-ribbon panel known as the Space Launch Broad Area Review (BAR) to be chartered by the Air Force and the National Reconnaissance Organization, to make recommendations to the government for improvements. Ironically, during the period in question, China had the best launch record of any spacefaring nation. The Space Launch BAR recommended that a combination of insight and oversight be used to ensure that the launcher manufacturers had paid proper attention to detail. The improvement in the U.S. launch record since this time could be used to argue that the recommendations worked, except that the "insight" and "oversight" is now precluded by the terms of a typical technology assistance agreement (TAA) for a foreign buyer as part of the U.S. licensing requirements. A foreign buyer of an American launcher is prevented from applying the same safety precautions to his procurement that the U.S. government has determined are necessary to ensure launch integrity.

October 27, 1999. A Proton rocket, carrying the Russian *Express A* communication satellite, failed. This was the second consecutive launch failure for the Proton with a Russian payload. Even though the problem—particles in the turbine of the second-stage engine—had been properly diagnosed after the first failure, the source of the particles was not properly identified and the corrective action implemented was incomplete. After the first failure on July 5, 1999, it was assumed that the foreign particles were introduced during propellant loading, and additional filters were added to the ground loading system.[9] Review after the second failure demonstrated, however, that the contamination was most likely to have been introduced during the turbo pump manufacturing process.[10]

The Russian Proton has been used to launch Western satellites since 1996, when it launched the *Astra 1F* for the Société Européenne des Satellites, a worldwide satellite services company headquartered in Luxembourg. The rocket is marketed by International Launch Services (ILS), a joint venture between Lockheed Martin, which provides services to the satellite manufacturer and owner, and Khrunichev of Russia, which manufactures the rocket. In addition, ILS offers the U.S. Atlas series of launch vehicles, and has recently developed contractual arrangements under which one launch vehicle can back up the other as part of a risk-management program.

The Proton examples cited here are good illustrations of the blatant silliness of the U.S. export-control process. Although the launch failures occurred during the launch of Russian payloads, Western Proton customers were understandably concerned. Before launching their satellites, they wanted assurance that the appropriate investigations had been performed and that all corrective actions had been implemented. Russians carried out the investigation with no American involvement. At the end of the investigation, the Russians briefed their American partners on their findings and recommendations, and they provided summary-level briefings to the insurance industry and satellite operators. During insurance and customer briefings in Europe, Americans were forbidden to be present due to export-control regulations—U.S. personnel might "slip" and give away valuable technology-transfer "know how." Any and all questions were encouraged in the European briefings so that prospective users could be satisfied that their launches were not at risk. When the briefing was given to a U.S. audience, Russians were prohibited from being present, so that questions extending beyond the officially approved briefing package could not be asked. The necessary information flow was essentially one way—from Russians to Americans—but the official concern is that American involvement could lead to an American providing information through the question-and-answer process that would allow a Russian to reach a conclusion that he could not reach on his own.

The purpose of rehashing these causes of launch failures is not simply to entertain, but to illustrate a point. None of these failures would have been prevented by applying rocket science. They may all be neatly cataloged in arcane engineering terms as "screw ups." The mistakes are not confined to a specific political system or ethnic group, and similar mistakes in the future cannot be prevented by technology transfer; they can be prevented by only hard work, attention to detail, independent review, and a commitment to excellence.

Surely one does not have to be a rocket scientist to prevent these errors. However, with the abundance of evidence that launch failures can be due to engineering ambiguity or human error, reliable space flight clearly requires the need to apply lessons learned from experience, follow proven processes, and employ a healthy independent review program. Allowing Americans to participate in these activities represents prudence, which is not at all the same thing as condoning technology transfer. Defining all such human interaction as "technical assistance" or "defense services," regardless of who is really assisting whom, creates the impression that export control is really misguided industrial protection in disguise; it certainly cannot facilitate a successful space program.

A Shift in Attitude

In a globalized world, and with the United States as a major proponent of globalization, linkages are being encouraged, free markets are being both expanded and promoted, and export sales are increasingly assumed as part of any manufacturing strategy. It is no different for the aerospace industry, and satellites in particular. Export sales are a critical piece of satellite sale projections. Subjecting satellite sales to the same controls and regulations as munitions, however, puts a considerable damper on the U.S. aerospace industry.

The crux of the problem is the ambiguity and obsolescence of current U.S. technology export rules. The Coordinating Committee on Multilateral Export Controls, known during its Cold War heyday simply as COCOM, long served as the basis for U.S. technology transfer. Created in 1949 by the United States and other NATO countries, COCOM proscribed countries such as the Soviet Union, other Warsaw Pact nations, and China. Under COCOM, member nations allowed other member nations to veto their export cases that required COCOM approval.

At the end of the Cold War, COCOM members, including the United States, recognized that COCOM's East–West focus was no longer appropriate, and efforts were initiated to develop a new multilateral agreement to replace COCOM, but maintain its intent. This evolution occurred through the early

1990s, culminating with the establishment of the Wassenaar Arrangement on Export Controls for Conventional Weapons and Dual-Use Goods and Technologies in 1995. There were thirty-three cofounding members of the Wassenaar Arrangement. Participating states maintain lists detailing dual-use goods and technologies, as well as other munitions items, but controls are administered on the basis of national discretion. That is where the United States has gone its own way.

The end of the Cold War changed the definition of national security in a number of ways. Militarily guarding against the Communist threat of the Soviets became a moot, though, in retrospect, sometimes easier mission. National security challenges proliferated horizontally in the military sense, as well as economically, environmentally, and in other areas related to transnational issues and globalization. The George H. W. Bush administration understood and responded to this change of situation. Addressing the increased role of economic competitiveness in national security, the president pocket-vetoed the Omnibus Export Amendments Act of 1990 and called to rationalize the munitions and commodity control list. All items on the COCOM dual-use list were to be removed from the U.S. munitions list, unless significant U.S. national security interests would be jeopardized by their sale abroad.

That it took six years for the directive to be fully implemented clearly indicates that the transition was neither easy nor clean. Technically, it was completed in March 1996, when the Clinton administration decided that the licensing lead for the export of commercial satellites should be taken from the often-erratic Department of State and given to the business-friendly Department of Commerce. The core of the export-licensing strategy is based on the COCOM premise of a U.S. munitions list (USML) that dictates what is subject to special regulation as munitions. The International Traffic in Arms Regulation comprises the implementation rules, policies, and definitions for the USML. Military satellites remained on the USML, and hence within the purview of State and the Department of Defense, but commercial satellites were removed. Many questions and ambiguities about licensing authority and the process remained. The transfer was more difficult in fact than in theory for two reasons: bureaucratic politics, and very different, often conflicting ideological perspectives between the organizations involved generally and some individual staff members specifically.

Remember that bureaucracies by their very nature do not actually produce or create anything. They regulate how others operate. Hence, self-perpetuation becomes a key organizational goal. When bureaucracies forfeit "turf," the goal of self-perpetuation can be perceived as threatened. The post–Cold War licensing shift inherently resulted in counterproductive, hurt feelings *cum* tacit abstention by some individuals and groups within the national security community, and parts of the State Department in particular, regarding licensing. In turn, this abstention inhibited a workable transition to a system that was to include the national security community.

The goals and processes of the State Department, the Commerce Department, and the Pentagon are fundamentally very different, though none is by any means a monolithic organization. In the State Department, export issues are handled primarily through the Office of Defense Trade Controls (ODTC), within the Political-Military Affairs Division under the undersecretary for arms control and international security affairs. In the ODTC and much of the Pentagon, national security has been narrowly defined in ideological and warfighting terms. Export licenses were, then, tools used with foreign governments to influence sometimes totally unrelated concerns. The Commerce Department, however, has seen and still sees a strong U.S. economy as an essential part of national security. Under that philosophy, issuing export licenses becomes routine unless good cause can be shown otherwise. Beyond the appropriate use of export licenses, differing views of technology transfer goals weighed heavily in decisions as well. One view was that stopping the flow of technology from one country to another is impossible, and, at best, only the rate at which technology flows can be controlled; the other view was that stopping the flow of technology is not only possible, but absolutely essential to U.S. security. The former view guided the Commerce Department, while the latter has been far more prevalent in the ODTC and Pentagon. What evolved between 1990 and 1999 was an environment of obstructionism from those seeking to stop the flow of technology, which subsequently led to frustration and sometimes circumvention by those in the industry seeking guidance. While the law ought to serve as the ultimate rule on the matter, it often has not.

The current U.S. export-control system is legislatively rooted in the Arms Export Control Act (AECA), enacted in 1976 to control the export of sensitive technology through licensing. But the State Department is not the only agency involved in export licensing, nor is the AECA the only law that regulates exports. According to John W. Douglass, president of the AIA, part of the problem lies with the legal framework:

> Several laws currently provide the statutory framework for export controls. In addition to the Export Administration Act and the Arms Export Control Act

> (AECA), the Trading with the Enemy Act and the International Emergency Economic Powers Act (IEEPA) are also used to control exports and international transactions. . . . The laws are not always mutually consistent, and ascribe primary administrative authority to different agencies.[11]

It has been strongly argued that all export and financial controls ought to be consolidated under a single act. So far, however, efforts in that regard have come to naught because of the different perspectives on the role of technology transfer. Some decision makers see current practices as too weak and therefore threatening national security. Others seeing them as too heavy-handed and threatening national security from an economic perspective.

For its part, the aerospace industry simply would like to see coherent and consistent rules. That is not likely to occur in the near future. As he left office in December 2004, Secretary of Commerce Kenneth Juster spoke about updating the expired Export Administration Act, saying that "he expects the Bush Administration to take 'a while' to form its position on the issue."[12] In the United States, the ODTC, with the assistance of the Pentagon, implements COCOM rules. The export-control process has always been an interagency process, including the State Department, the Pentagon, and the Commerce Department. But who plays what role, and who has authority for what, is not always clear.

In 1995, after the explosion of the *Apstar-2* satellite in China, Hughes Electronics, the manufacturer, met regularly with Commerce Department officials, because Commerce had granted the initial export license. The meetings concerned what Hughes could or could not say to the Chinese regarding what they suspected was the reason for the failure. Hughes worked within the parameters given to it by Commerce and had technical exchanges with the Chinese cleared by Pentagon officials. But as it turned out, Commerce had no authority to grant approvals; State had jurisdiction for issues related to the failed mechanism, because it was considered related to the launch vehicle, not the satellite. Hughes was considered in violation of government regulations. It had sought and received approval for everything it had done, but from the wrong department.[13] Some Hughes employees later admitted during the Cox Committee investigation that while the licensing parameters were ambiguous, they knew, or suspected, that State was actually the correct approving authority.

It is also often unclear in what instances Pentagon monitors are required to attend meetings between satellite owners and launch officials. There were times after 1996 when the aerospace industry was told by one individual at State or the Pentagon that monitors were not required at all meetings involving commercial satellites, while another official stated that they were mandatory. Additionally, the Pentagon occasionally declined to send monitors without explaining whether it was because monitors were unnecessary or because the Pentagon substantively objected to the meeting. Indeed, who to ask and who was an authoritative source for advice was even difficult to decipher. Consequently, the industry goal evolved to not just getting questions answered, but getting answers that would expedite commercial goals. Managers get rewarded for results, not excuses. Getting a negative answer was not necessarily seen as either definitive or in accordance with precedent. Without question, the industry pushed the envelope of what was allowed more than once, and this obviously left its actions open to challenge later by Congress and the media during congressional hearings.

The Cox Report

Since 1999, the U.S. export-control system has been driven by both legislation and decision making that directly resulted from the Cox Committee report.[14] Part of the thousand-page,[15] five-volume, classified report completed in January 1999 detailing the theft and acquisition of secret U.S. technologies by China was declassified and released in May 1999. The glossy-covered, three-volume unclassified set was more like the Soviet Military Power series produced by the Reagan Administration than like typical committee reports. Its importance, staff members stated, justified the divergence from the drab green covers usually used.

The findings were as superficial as the cover. According to the report, American nuclear weapons designs were jeopardized, Chinese missile capabilities were improved, and American security was hanging by a thread. Andrea Mitchell, interviewing Christopher Cox for the evening news, asked the kind of nebulous yet frightening question that became typical of the media coverage of the Cox Report: "Congressman, do you sleep a little more uneasily at night as a result of all that you know now?"[16] After the 2000 election, the near panic over national security concerns vanished, and those most loudly shouting that the sky was falling quickly disavowed involvement and called for efforts to unravel the mess created by the Cox Committee recommendations. This suggests that the melodrama was more a function of politics than technology.

A group of Washington "insiders" with strong anti-China views, including congressional staffers, think tank analysts, and agency personnel, began meeting in the mid-1990s. They called themselves the Blue Team, referencing the color assigned to the good guys in military war games. They supported a confrontational policy toward China and military and diplomatic support for Taiwan. They looked for opportunities to portray China as an imminent threat. Many were, and are, also neoconservatives and avid supporters of missile defense. This is not unimportant. Showing that China had increased its missile capabilities bolstered the case for missile defense that was then being strongly pressed after the 1998 report of the Commission to Assess the Ballistic Missile Threat to the United States, headed by Donald Rumsfeld. After Clinton survived first the Whitewater scandal and then impeachment, several issues became stand-in targets for assaulting Clinton—his China policy being one of them. The Blue Team lit the spark surrounding the allegations of technology theft and illegal transfer, and Republicans in Congress were more than happy to throw gas on the fire, linking it to other scandals and revelations involving campaign financing and corporate greed, all while gathering support in the perpetual quest for missile defense.

The media played an important role in sounding the alarm about both charges investigated by the Cox Committee as well. Initially, media sympathetic to and sometimes associated with the Blue Team, such as the *Washington Times*, covered the story. It did not take long, however, for outlets like the *New York Times* and others to follow suit,[17] and in sensational form. On April 4, 1998, the *New York Times* ran a front-page story by its Washington-based investigative reporter Jeff Gerth, working with Raymond Bonner, entitled "Companies Are Investigated for Aid to China on Rockets." The story focused on the federal grand jury investigation into whether two American companies illegally gave China space expertise that significantly advanced its ballistic missile program. The companies were Loral Space and Communications and Hughes Electronics. The article stated that the inquiry stemmed from a 1996 launch accident (that of *Intelsat 708*) in which a Chinese rocket carrying a satellite built by Loral exploded shortly after liftoff. After the accident, both companies independently reviewed the failure and reported their assessments to engineers at the Chinese Academy of Launch Technology (CALT). In doing so, the article stated, the companies may have provided "crucial assistance" to China for guidance systems used in ballistic missiles. Also stated in the article was that "Loral has numerous business deals with China and close ties to the White House. Its chairman and chief executive, Bernard L. Schwartz, was the largest personal contributor to the Democratic National Committee last year," clearly implying a trade of favors for money.

While the grand jury investigation was under way, the article continued, President Clinton "quietly approved" permission for Loral to launch another satellite on a Chinese rocket. The implication that the approval was done on the sly certainly made it appear to be a nefarious act. The permission was regarding a special waiver required for launching U.S. satellites on Chinese rockets, put into effect after the Tiananmen Square massacre in 1989. The requirement for such a waiver was instituted because of concerns about Chinese human rights violations, not technology transfer. The article pointed out that "Congress must be told of each waiver [permission]. Thus far, Presidents Bush and Clinton have issued 11 waivers for satellite launchings." Hence, Clinton's quiet approval included notification to 535 members of Congress, and was one in a line of eleven approvals that started during the prior Bush administration. It would have been more newsworthy had Clinton not approved the waiver, as all previous waiver requests had been approved.

Gerth followed that article with another on April 13, again running on the front page. In this one, "U.S. Business Role in Policy on China Is Under Question," politics and scandal began to have a more prominent role through more references to party donations. It began by stating that "Loral and Hughes tilted towards the Democratic Party, giving $2.5 million to Democratic candidates and causes, and $1 million to the Republicans." Then the article alluded to a connection between party support and action: "the President's desire to limit the spread of missile technology was balanced against the commercial interests of powerful American businesses, many of which were White House allies and substantial supporters of the Democratic Party." Clearly, the insinuation was that the president gave more weight to commercial interests than to U.S. security in his decision.

One day later, on April 14, 1998, the *Wall Street Journal* ran an editorial, referring directly to Gerth's article of the previous day. It was even more politically sensational, as well as factually incorrect:

> President Clinton approved the transfer of missile guidance technology to China at the behest of the largest personal contributor to the Democratic Party. He granted the needed waiver despite an ongoing Department of Justice criminal investigation of the same company's earlier transfer of similar technology: a Pentagon study concluding that in the earlier episode "United States national security has been harmed."
>
> That is the essence of a report yesterday by Jeff Gerth of *The New York Times* concerning satellite launch technology provided by Loral Space and Communications and Hughes Electronics, a subsidiary of General Motors....
>
> Expectations that a President will resolve such issues on the merits, though, have been deeply muddied by the ongoing controversy over the Clinton contributions scandal and Chinese money.[18]

Gerth's stories were enough to spawn the first of three "revelations," on which the entire Cox Committee saga was based, implying exposure and surprise. The *New York Times* campaign finance reform reporter, John M. Broder, wrote on May 20, 1998, "There is a difference in the tone of responses to the revelations of the [administration's] dealing with China. Absent are the president's accusers and the cries that the accusations are the product of mere partisan politics."[19] The new revelations were clearly implied as more serious.

The first of the revelations was that satellite manufacturers want their satellites to be launched successfully. The accompanying insinuation was that aerospace companies' greed was stronger than their concern for national security; hence, the illegal technology transfers. It was suggested that the administration allowed and abetted the nefarious and illegal activity after having received campaign contributions. The second and third "revelations" were quick to follow: rockets and missiles share technologies, and China was involved in espionage. If those revelations were really a shock to anyone in Washington, home of the largest intelligence (espionage) operation in the world, then there were larger national security problems to be addressed than those undertaken by the Cox Committee. In May 1999, after the declassified version of the Cox Report was released, Leon Hadar commented on the insincerity of the "revelations" aspect of the Cox Committee investigations, noting that, like the French police inspector in *Casablanca*, the committee was, "shocked, shocked!" to discover illicit dealings that were apparent to any but the most obtuse observers.[20] Hadar's views were not newsworthy in the United States, and he did not win a Pulitzer for his work on this issue. Gerth's views were prevalent, and he did win a Pulitzer.

Later media analysis pointed out that the Cox Report was approved on a bipartisan basis.[21] Bipartisan approval was interpreted as evidence that partisanship had no role in either the investigation or the findings. Statements made after the report was released, however, suggest that the bipartisanship was only an image. In a post–committee report speech to the Republic National Committee on January 22, 1999, Christopher Cox denounced the Clinton administration for cozying up to Beijing and accused President Clinton of giving China's leaders "the full Lewinsky." The ranking Democrat on the committee, Norm Dicks (D-Wash.), said after completion that the report represented a "worst case assessment." For the most part though, Democrats said very little. They worked diligently to make China a nonissue in the 2000 presidential election and avoid further scandal. The Democratic tactic was basically to keep silent or, when necessary, agree with whatever new revelation or allegation was made, to preempt any opportunity for Republicans to charge that the Democrats were defending either China or corporate greed.

Rebuttals did, however, quickly appear—mostly from academics and technical experts beyond the reach of government influence. The most detailed of those was a ninety-nine-page analysis by a panel of scholars from Stanford University. The panel's expert on Chinese governance and policy was Alastair Iain Johnston, a Harvard professor who was a visiting scholar at Stanford. The nuclear weapons section of the report was by Wolfgang K. H. Panofsky, a former director of the Stanford High Energy Physics Laboratory; the Chinese arms control section was by Marco Di Capua, a Lawrence Livermore physicist who had served at the U.S. Embassy in Beijing from 1993 to 1997; and the section on China's acquisition of U.S. missile technology was by Lewis R. Franklin, a career intelligence expert on Sino-Soviet missile and space research who was a visiting scholar at Stanford. The report stated that "important and relevant facts [in the Cox Report] are wrong and a number of conclusions are, in our view, unwarranted."[22] Others agreed, and they were not all woolly-headed liberal academics. Jonathan Pollack, a China and defense specialist at the RAND Corporation, also spoke out. He called the report an "unbelievable rush to judgment," expressing particular concern about the nuclear theft assessment being especially weak. Pollack was quoted as saying, "If this were a Ph.D. thesis at RAND, I would flunk it."[23] Commentary from the expert communities holding "opposing views" went on, including those published by Richard L. Garwin in *Arms Control Today* and Lars-Erik Nelson in the *New York Review of Books*.[24]

Just before the Cox Report was published, the Senate Select Committee on Intelligence released (with little fanfare) its own report on the topic.[25] While critical of the Clinton administration, the report stopped short of concluding that China was close to altering the strategic nuclear balance. An interagency "damage assessment" team created at the request of the Cox Committee to investigate the implications of China's nuclear espionage concluded that significant deficiencies remained in China's weapons program and that its aggressive collection effort had not resulted in any apparent modernization of its deployed strategic force or any nuclear weapons deployment.[26] China's 20 ICBMs remained dwarfed by the 12,000 weapons in the U.S. nuclear triad.[27] The sky was not falling.

Even before its release, the Cox Report was sensationalized. Republicans claimed that the delay between the report's completion and its declassified release stemmed from the Democrats buying time to prepare a counterattack, and the Clinton administration trying to hide the damning evidence regarding the passing of ballistic missile technology to Communist China by American satellite manufacturers. Nevertheless, its findings were anticipated (read: leaked) and subsequently were reflected in both legislation and decision making. The Strom

Thurmond National Defense Authorization Act of 1999 directed that on March 15, 1999, control of U.S. communication satellites for export would be given back to the Department of State from the Department of Commerce. Further, when the Cox Report was finally issued, it contained a series of recommendations for stopping the spread of sensitive technology. Some of the recommendations were not well thought out, resulting in unintended consequences. The problems they caused were exacerbated when the State Department not only embraced them, but expanded on them as well. Within the State Department, the Office of the Undersecretary of State for Arms Control and International Security, headed by John Bolton from 2001 to 2005, was considered particularly zealous on export controls. The Strom Thurmond Act apparently intended to exclude European and Japanese launchers from the most draconian of the oversight rules, but the State Department included them. The act says that some of the most intrusive aspects of the controls do not apply "to the export of a satellite or related items for launch in, or by nations of, a country that is a member of the North Atlantic Treaty Organization or that is a major non-NATO ally of the United States."[28] But the new law did not modify the International Traffic in Arms Regulations already on the books, which make no special exceptions for U.S. allies. The State Department reserved the right to apply the controls to everyday transactions with allies and exercised that right.

Additionally, the State Department unilaterally decided that export licenses were required to return hardware to its country of origin. If a part arrived in the United States from Europe damaged, a license had to be obtained to send it back for repair. How U.S. technology was in danger of being transferred in such a scenario is uncertain. And whereas the Commerce Department had deemed many items, such as screws, as not requiring licenses, State refused to recognize the "no license required" determinations. Not only were the regulations ambiguous, but they now extended over a much broader scope of items, and State had neither generally nor specifically trained personnel to deal with the paperwork. At one point, license applications literally had to be physically carried from the State Department to the Pentagon because their computer systems could not talk to each other. A self-fulfilling prophecy for an out-of-control situation was created.

Control What? How?

Changes to the export-licensing system in 1999 were theoretically intended to stop rogue states and potentially hostile countries—at that time, read

China—from acquiring sensitive U.S. technology, using USML definitions of "sensitive." Most hostile parties, of course, do not apply for licenses. Furthermore, since the end of the Cold War, the United States has not had exclusive control over much of the technology in question. The USML criteria for defining what constitutes sensitive technology are arbitrary and almost useless. Sometimes the criteria definitions are ambiguous, sometimes they are rigid, and sometimes they are outdated because of the commercial availability of the "sensitive" technology they are intended to control. This means that stopping the flow of technology from the United States does not necessarily stop China, or anyone else, from acquiring it; it only stops it from coming from the United States. Whether the United States approves or not, countries can buy what they consider to be vital technology elsewhere, with perhaps fewer safeguards than the United States requires—and other countries are eager to sell it to them.

Many countries have commercial contracts with China to acquire both hardware and knowledge. Italy's Telespazio contracted with China for the sale of image-processing computers, software, and training personnel. China has contracts for commercial radar satellite data, required for all-weather, day and night monitoring of military activities from multiple foreign vendors, including Japan and Canada. The 1997 contract for Canadian Radarsat data includes training imagery analysts. Russia is working with China on surveillance systems, propulsion, joint design, scientific personnel exchanges, space systems testing, and satellite navigation. China has hired Russian advisers in the areas of ballistic missile guidance and control technology. France, Kiribati, and Chile have agreements with China regarding the joint use of ground stations for data relay satellites, necessary for tracking satellites passing overhead.[29] The United States cannot hope to persuade other countries to forgo lucrative contracts. But more important, as the United States inhibits its own technology exports, other countries eagerly pick up the slack created by U.S. market withdrawal. The technology transfer goals of the Cox Report recommendations are unattainable by the means provided. Meanwhile, the unintended consequences of the system subject a key area of its industry to regulations that are arbitrary and capricious at best and detrimental to U.S. security at worst.

Current technology transfer rules require assurance that both hardware and knowledge cannot be passed on. How this can possibly be accomplished beyond a reasonable degree is left unstated. Difficulties in trying to safeguard specific technologies arise not so much in hardware, but in intrinsic knowledge about the hardware, which administration officials fear can be passed from the supplier. Arms control officers sometimes privately admit that it is impossible to stop or even really define what is or is not acceptable for "know-how" transfer, but the requirement to try remains. The situation is further complicated because the United States can control only what it owns, and it owns neither all the hardware nor the know-how that other countries, including China, want to acquire.

The United States has returned to the COCOM format, characterized by rules that may be desirable in theory but not enforceable in fact. Consequently, the U.S. aerospace industry is left wandering in a bureaucratic wonderland at a time when international economic competition is not theoretical, but very real. Further, nonproliferation and security goals have actually become harder to achieve because it is impossible to design effective enforcement procedures for controlling know-how in an increasingly globalized world. Physics and engineering are not areas under the exclusive control of the United States. Subsequently, procedures to limit the sharing or expansion of knowledge end up both ill-defined and impossible to execute.

All satellite technologies again became part of the USML. It is unclear how far the scope of that mandate dips into subsystems and components. It is clear, however, that the reach of the new regulations can be broad. Testifying before the Senate just after the change in regulations in 1999, AIA president John Douglass told a story about a small U.S. company that got a contract to sell airplane landing-gear knobs in Europe. The knobs are pulled down to lower the aircraft landing gear, and pushed up to raise the gear. When the Europeans came to inspect the knob that they had paid for, the State Department made the U.S. manufacturer put a bag over the product they were trying to sell, and the license was delayed.[30] In May 1999, the Italian military, supporting NATO's Operation Allied Force with a search-and-rescue operation in the Adriatic Sea, requested a license from the United States to purchase flares, which was inexplicably denied.[31] The reach of the licensing process has been long and unpredictable.

Time is money in business. The State Department claims that it issues licenses within thirty days of submission. A few caveats, however, are important. First, thirty days is measured by working days, not calendar days. And the license for a satellite itself is often not the problem. The problem lies in obtaining TAAs required to hold marketing discussions and to exchange basic technical information with insurers and launch service providers. Exporting a single telecommunication satellite requires several TAAs. Congress must also be notified before a license is issued, and State will not send up a license for review when Congress is out of session. This cumbersome process through State is one of the reasons that licensing was moved out of that department in the first place. Typically, buyers working with European countries can resolve similar issues in twenty-four hours.

Unintended Consequences

In mere anticipation of the Cox Committee findings, a U.S. company was for the first time actually denied the requisite license needed to export a satellite to China. In a surprise February 1999 decision, the Department of Commerce informed Hughes Electronics that it would reject the company's export license request for the commercial Asia Pacific Mobile Telecommunications (APMT) satellite, intended for a Singapore-based consortium with financial links to the Chinese military. There is little doubt that the Commerce Department was cowering in the face of the politically charged atmosphere in Washington surrounding the Cox Committee, even before the report's declassification and release. Although Clinton administration officials stressed that denying the AMPT license was a one-time event, not a fundamental policy shift,[32] the chilling effect of the decision has been tangible.[33] Increasing the chill is that the grounds for the rejection were never fully explained to the public. Whether it was because of the end-use of the satellites, concerns about technology transfer during the execution of the contract, the potential for technology transfer during launch, or politics remains speculative.

Not long after the APMT debacle, another Chinese satellite, this one a Loral standard "bent pipe" model sold to China Telecommunications Broadcast Satellite Corporation, was held up. First, the Pentagon declined to send a government monitor to a technical meeting. Then, in a typical head-scratching conundrum, although Loral obtained an export license for the ground-control software needed by the Chinese to control the satellite in orbit, the Pentagon barred Loral from installing it at the Chinese ground station.[34] Whether bureaucratic inability to respond or politics was behind the impediments is unclear, but the result is the same.

In spite of the Clinton administration's assurances, then, it is now clear that a fundamental policy shift had in fact occurred. Satellites that had been under construction for Chinese entities were held on the ground (*Chinasat 8*), or ownership was transferred to allow launch (*Apstar V*, now *Telstar 18*). All Chinese satellites are now being built indigenously or by Alcatel Space, a French company.

The post–Cox Committee regulatory environment in the United States has certainly been a windfall for European and Japanese aerospace firms, but not without some pain on their part, too. Initial European jubilation over the increased business was short-lived. Europe quickly moved from concern about the proposed changes to U.S. export regulations to frustration, bordering on outrage, with the new technology export policy. Under the new export rules, components that had been approved and contractually committed years earlier for the next-generation Intelsat satellites, including such "sensitive" technology as screws, were delayed in shipment to Paris-based Alcatel Space. Intelsat threatened to move its headquarters outside the United States due to "unfriendly" signals it perceived from Congress.[35] German officials considered appealing the new policy to the World Trade Organization.[36] The complaints and skepticism about the dependability of the United States as a partner continue. Export controls are not the only area in which the U.S. image as a reliable partner has suffered; it is just another area that contributes to the perception of an overall decline in American space leadership.

Meanwhile, European firms quickly made headway into the Asian aerospace market, particularly in China, avoiding the long tentacles of U.S. restrictions. Almost immediately after the Strom Thurmond Act took effect, Germany's DaimlerChrysler Aerospace decided to divest itself of U.S. subcontractors. The rationale was that it could not afford to sit idly while the State Department decided its business fate with costly licensing delays. Even Telesat Canada, a longtime loyal customer of U.S. satellites, decided in 2003 to buy from France's Astrium rather than risk licensing issues with the United States.

Domestically, in 2002, Loral, one of the two companies investigated by the Cox Committee for alleged illegal technology transfer, agreed to pay a $14 million civil fine for having passed missile technology to China, while neither admitting nor denying the charges. Hughes Electronics (now Boeing), the other company involved, agreed in 2003 to a $32 million fine. Both cases began as criminal investigations. They ended up in civil court after the government could not prove its case. Why, then, would the companies agree to pay the fines and settle?

First, having to deal daily with the State Department puts the fate of a firm's business in State's hands. Fighting or even criticizing those who control one's fate is not a smart business move. Almost certainly, if some of the denied export-licensing requests were appealed to a civil court, a strong case could be made for arbitrary and capricious decision making. But what company can afford to challenge the system in court without risking retribution? The choice seems to be between making a point and potentially finding oneself out of business, and keeping quiet and staying alive.[37] Second, court cases with the government do not look good to stockholders. The U.S. aerospace industry is still healthy, but it is certainly stressed. It is endeavoring to work quietly within the system, whatever it is, by obtaining working definitions to portions of the new regulations in small increments. Still, sometimes paying fines and working within the system is not enough. On July 15, 2003, Loral declared bankruptcy.

Fixing the Fix

There have been ad hoc efforts to "fix the fix," or at least make the system workable within the parameters of the USML and the Department of State. Initially, waivers were the best that could be offered to keep business moving. In late summer 1999, the main antenna for the new Intelsat IX series of satellites, manufactured by DaimlerChrylser Aerospace in Germany, was shipped to prime contractor Loral in California. Unfortunately, it was improperly packed in its shipping container and was damaged in transit. Loral was first told by State that it needed an export license to ship it back to its country of origin for repair. This could have resulted in a delay of three to six months, with late delivery penalties accruing on the satellite at the rate of $70,000 a day. Intelsat was outraged: it wanted its satellite on time, not late fees. In this case, the State Department issued a license in just under a month as a no-brainer. This was the good news, and such licenses have been issued with increasing frequency. The bad news is that over a month, considerable late penalties still accrue.

Also in November 1999, Congress included provisions in the omnibus appropriations bill directing the Department of State to expedite considerations of satellite exports to NATO allies and major non-NATO allies, such as Australia, Israel, and Japan. Section 1309 of HR 3427 states that expedited treatment should be given to "time critical" projects, submission of contract bids to foreign firms, reexport of "unimproved materials, products or data," and data required to obtain launch insurance.[38] Unfortunately, changes have been slow and did not remove the sting from having been, as allies, subjected to the same treatment as potential adversaries in the first place.

Politicians quickly recognized that the post–Cox Committee export license "fixes" were counterproductive. By May 1999, with the official release of the Cox Report, indications of discontent had already begun appearing in the press: "Members of Congress are clearly growing concerned about the potential damage to the U.S. satellite industry and some are blaming overzealous government bureaucrats for the slowdown of export license decisions that began March 15 when authority was transferred to the State

Department from the Commerce Department."[39] James Sensenbrenner (R-Wis.), chairman of the House Science Committee, who had voted with the majority to revert licensing authority to the State Department, addressed the issue before an American Institute of Aeronautics and Astronautics audience in May 1999. He said that when his staffers were crafting a bill dealing with commercial space activity, they encountered State Department officials "who did not understand the laws they were enforcing, the technologies they were regulating, or the international nature of high-tech industries."[40] Aerospace industry officials were likely not surprised. Sensenbrenner and other members of Congress maintained that timing had kept them from getting their concerns about the State Department's attitude considered during the Cox Committee deliberations. Less than two months after the shift of licensing responsibilities to State, Sensenbrenner said he was in favor of regulating satellites from the Commerce Department and that he was "floating a trial balloon" about Congress potentially returning satellite-export authority there.[41] That has not happened.

Representative Dana Rohrabacher (R-Calif.) is one of the most anti-China members of Congress. According to his Web site, "Rohrabacher is one of the committee's most outspoken advocates of human rights and democracy around the world. He is the leader each year in the effort to deny Normal Trade Relations status to China, a role he will continue to play until that communist government respects the human rights of its citizens, refrains from threatening its neighbors and lowers tariffs on American products."[42] As such, he was also among the most histrionic during various committee hearings on the issue of technology transfer to China. At one of the many other congressional hearings on alleged missile technology transfers to China spawned by the Cox Committee, a witness suggested that assertions that commercial satellite launches had facilitated improvements to the Chinese missile arsenal were based on "half-facts that have been distorted and sensationalized." Rohrabacher angrily and indignantly shot back, "The blueprints for all these changes were not left under their [China's] pillow by the tooth fairy."[43] There had never been any allegations of blueprints being involved. Regardless, even Rohrabacher has now changed his mind. Apparently, unemployed aerospace workers in his California district have caused him to rethink his position.

All fingers began to point at the Department of State as responsible for the situation. But State eventually tired of being Congress's excuse for why the licensing process had suddenly and seemingly inexplicably almost hit gridlock, with customers threatening to turn away, or actually doing so, in droves. At a June 24, 1999, hearing before the Senate Committee on Foreign Relations, Eric

Newsom, assistant secretary of state for political military affairs, stated, "All of these measures are being implemented very strictly—and *aggressively*, as recommended by the House Select Committee—through our munitions license process."[44] The surprise that some members of Congress expressed about the State Department attitude was puzzling. State had not wavered during hearings about what criteria it would use for export licenses, and at the time, it had been music to congressional ears. The blame game revved up quickly, and then faded away just as quickly, leaving the unworkable system as standard operating procedure.

Again, the media had been an active partner in provoking much of the upheaval. Then, on September 7, 1999, the *New York Times* ran a front-page story by its long-time and well-respected science writer, William Broad, that caused more agitation in newsrooms than it did in the public. Entitled "Spies vs. Sweat: The Debate over China's Nuclear Advance," the article was what some called not-quite-a-retraction, but certainly a "5,000 word correction" to earlier articles.[45] Broad concluded that the Cox Report overstated the link between alleged espionage and China's advances and that the Federal Bureau of Investigation (FBI) rushed to judgment when it focused on Wen Ho Lee. By that time, however, the Cox Report findings had been accepted as truth. Nobody cared about corrections.

Throughout the Cox Committee investigation, businesses basically remained mute, apparently intimidated by the degree of hostility coming from Washington. But Republicans, traditional allies of big business, were eventually visited by representatives from Intel, Sun Microsystems, and Unisys Corporation to talk about computer exports to the burgeoning China market. Representatives from other sectors followed. Subsequently, in July 1999, Congress approved extending most-favored-nation (MFN) status to China, allowing business as usual to continue with China in most commercial areas. Conferring MFN status means that imports from that country receive favorable rates of duty. So trade with China in other areas continues unimpeded. The aerospace industry, it seems, assumed almost a symbolic role through which individuals and groups could castigate China or the Clinton administration. However, in 2003, the electronics industry and high-performance computer companies seeking to do business in China began to feel the pinch of U.S. export-control laws as well.[46]

If one goes by the materials that the State Department publishes on the Web, export-control regulations now seem almost simple and easy. There is even a site that includes questions and answers about exports, and then quizzes the reader on what has been learned.[47] But it is neither simple nor easy;

on the contrary, it is broken. George Abbey and Neal Lane list export-control policy as the number-one challenge to the future of the United States in space.[48] Representative Dave Weldon (R-Fla.) joined a long line of those urging a "sensible export licensing process that will happen in a fair and timely manner."[49] Why has the Bush administration not stepped in to stop the silliness? The administration has been divided on the issue. The tension lies between the free-trade wing of the Republican Party and the Cold Warriors who view China as the next competitor of the United States. The latter include the Blue Team and its associates, who instigated the Cox Committee actions.

The effects of the Cox Committee are still very much felt, though as of 2005, some glacial and incremental changes have occurred. On September 23, 2002, government regulations were published in the *Federal Register* that allowed "some" items that routinely fly on satellites to be licensed by the Department of Commerce and, unless they breach certain criteria, not require a license.[50] Positive developments have been slow because the regulations are ambiguous. Solar cells and microwave components can be licensed by Commerce, but what might constitute a microwave component is left undefined. At least one aerospace company has sent lists of microwave components to Commerce to confirm that the department has jurisdiction over them. When Commerce agrees, the list is then sent to State and the Pentagon for their confirmation as well. So far, neither State nor the Pentagon has responded; the industry requests for confirmation of jurisdiction seem to have disappeared into the same black hole as have many of the license applications. Having been burned more than once by ambiguous regulations, industry officials have been reluctant to assume any risk in that regard. Further, while solar cells are apparently approved for Commerce licensing, batteries are not. It is not at all clear on what basis these licensing transfer decisions are being made. Ambiguity and arbitrariness continue to be the norm.

Security Ramifications

Since the fallout from the Cox Report changed U.S. export-control laws, countries and companies increasingly choose not to work with the United States for fear of being caught up in the quagmire of its export-control system. Fewer satellites have been ordered annually since 2001 than earlier projected, which affected not only the satellite industry, but the launch industry as well. Launch manufacturers lost interest in investing their own money in the Evolved Expendable Launch Vehicle program, resulting in the high government overruns discussed in chapter 4. With less business as result of a confluence of several factors, the long-standing leadership role of American companies becomes jeopardized. Because the military increasingly relies on commercial satellites, that leadership role is required to maintaining U.S. space superiority as well.

The loss of leadership in the aerospace industry through decreased exports might well be considered an acceptable trade-off if it heightened U.S. and world security by keeping sensitive technology from those that might put it to nefarious use. But this is a false premise. If a country wants technology, it will obtain it, as North Korea and Iran proved by obtaining nuclear capabilities. The United States may have been the first to launch a geosynchronous communication satellite nearly forty years ago, but that achievement was based on scientific and engineering knowledge that had been established over hundreds of years by people of all nationalities, only a few of whom were American. Much of the engineering developments of the past forty years are now in the public domain. Possible metaphors abound, but locking the barn door after the horse has escaped is a perfect description of U.S. efforts to control the use of space technology.

Regulators purport to defend national security, but they have never articulated their specific technology concerns. Is the United States concerned about the use of space launch vehicles as offensive missiles? This would reverse a consistent historical pattern. Most current space launch vehicles were offensive ballistic missiles first. Is it a concern about the proliferation of communication satellite manufacturing and design technology? If so, then it seems odd to create incentives to purchase less regulated, non-U.S. satellites. Perhaps the concern is about the possible use of space communications by foreigners with interests opposed to those of the United States, including terrorists. If so, then we should also stop the proliferation of Internet cafés and sales of cell phones.

The world is not without risks, and those risks should be recognized and addressed. But empowering bureaucrats to regulate those risks based on unexamined assumptions, unstated objectives, and without any means of measuring the results is irresponsible and counterproductive.

The technology export process has been broken since the end of the Cold War. Not only did a more ambiguous world present new questions regarding what to control, how, and from whom, but the answers to the questions got stuck in a bureaucratic quagmire of epic proportions. With the injection of politics into the process during the Clinton administration, the situation worsened.

The aerospace industry as a whole may prosper, but the fate of all sectors will not be the same. The buckets of money being poured into missile defense efforts may keep some portions healthy, but not communication satellites. The United States lost its dominance in the launch industry because of cumbersome regulations. With the political debacle following the Cox Committee report and the legislative changes to the export process, the satellite industry seems destined to follow the same path.

The findings of the Cox Committee have become urban legend. I have lost count of the number of times I have been asked to answer questions from the media based on the findings of the Cox Committee, only to have to dispel false notions. The committee provided drama at the ultimate expense of national security, from an economic perspective, from a perspective of potentially losing leadership in a field critical to the military, and through denying technology to countries and thus pushing them to other suppliers. As pointed out in chapter 3, while recognizing the need, the Bush administration has not followed through on rationalizing the export-control process. Politics—this time in the Republican Party—continue to place the American aerospace industry at risk. Meanwhile, the European and Asian aerospace industries, including in China, are continuing to advance. While their technology is still not as good as that of the United States, in some cases it is not far behind, and it is more easily accessible. The aerospace industry, on which the U.S. military and economy depend, thus becomes threatened. We are committing strategic suicide.

The Ambitions of Europe

SEVEN

Security and defence are inconceivable
in the 21st century without appropriate,
reliable space-based capabilities.
—Assembly of the Western European Union

All countries use space in some way, but far fewer are really space players. Among the developed countries, the primary space actors are European countries—both individually and collectively through the European Space Agency and activities of the European Union—Russia, Japan, and Canada. Russia is still a major player, if only because of its legacy. Today, scarce government resources have pushed the Russian space program into being perhaps the most entrepreneurial in the world, often to the dismay of Washington. In addition to national programs, European countries adopted a "hang together or hang separately" attitude in the 1960s that continues in many civilian areas and is being considered for extension into others. European integration has been highly successful in many areas of economic interest, from adopting the euro as a common currency to developing the Airbus to rival U.S.-made airplanes. In areas of politics and national defense, however, integration has been slower. Space-related national defense activity, for example, has been a purely national purview, with joint activities primarily found in the scientific and commercial areas. That is changing. Japan's program is ambitious, though economically constrained and ultimately culture-bound. Japan's bureaucracy-led, consensus decision-making style results in a risk-averse, glacially moving process that makes advancement painful at best. Canada provides a model for countries with limited budgets that want to engage in space activities. Canadian efforts were originally pragmatic, focused on areas with direct and near-term payoffs for the Canadian people, such as robotics and telecommunications, and were highly successful in that regard. But even with that approach, Canada has struggled. The high cost of space operations renders space programs vulnerable when they compete with other national funding priorities. Overall, however, the European experience perhaps best illustrates the issues being raised and actions being considered by developed countries with space capabilities.

Initially, European countries felt it imperative to be involved in space for the technology it yielded. Technological development led to industrialization, and industrialization led to economic growth. While still an important motivation, technological development has taken on a political dimension as well: it is viewed as an indicator of political power; according to Stefano Silvestri, "Space security applications are directly linked to the role of Europe in the world. The negotiations between the U.S. and the EU on the Galileo system clearly confirm this."[1] So while space activity for the sake of "prestige" is sometimes scoffed at, prestige that translates into increased political status and gives countries a voice in regional and even global affairs that they might not otherwise enjoy is a new form of geopolitics.

Hang Together or Hang Separately

Since the early days of the American–Soviet space race, European countries have been concerned about being left behind in a race not only to develop the technology to get to the moon, but to spur industrialization, and thus economic growth, on earth.[2] Levels of concern varied, usually according to the ability to do something about it, resulting in unequal national efforts. France and Germany developed their own relatively robust national space programs, while other countries struggled due to technical constraints or, as in Great Britain, organizational mayhem. While Great Britain worked on the Blue Streak and Black Knight missiles after World War II, it was continually frustrated by bureaucratic politics. At one time, responsibility for "space" was assigned to no less than nine ministries, sometimes simultaneously.

Germany's interest in space has been for scientific research, with the ultimate hope of yielding industrial and commercial payoffs. But while Germany's World War II legacy with its V-2 rockets provided the United States and the Soviet Union with considerable scientific knowledge after the war, it also resulted in domestic reluctance to become engaged in independent launch efforts, especially in assuming a leadership role. France had no such qualms; it was originally interested in developing an independent launch vehicle as a military capability, with civilian and commercial launches viewed as a potential side benefit. France has always coveted the role of European space leader.

The successful launch of the Diamant-A rocket in 1965 made France the third country in the world with an independent launch capability, behind the United States and the Soviet Union. Even then, France felt it imperative that Europe as a whole move away from dependence on the United States. However, in conjunction with initial European integration efforts in the 1960s, Western European countries soon recognized that none of them had the resources or the talent to establish themselves as independent space players. Individually or collectively, European countries worked with NASA throughout the 1960s, NASA's golden age of cooperation. They flew experiments on NASA missions and participated in ground-based experiments, scientific and technical exchanges, and even satellite launches. Meanwhile, NASA launched the first two European scientific satellites, *ESRO II* (*Iris*) in May 1968 and *ESRO I* (*Aurorae*) in October 1968, on Scout rockets, free of charge. By the end of 1969, NASA had launched nine European spacecraft from France, West Germany, Italy, and the United Kingdom. As Reimar Luest relates,

> Of course, every single European state and later Europe as a whole, gladly accepted the help offered by the U.S. through NASA for their first steps into space, since for the countries initially approached, such as the U.K., France, the Federal Republic of Germany, Italy, it was the most efficient way to build up capabilities and capacity in space science and technology and obtain in a reasonably short time scientific results from space missions.[3]

European countries recognized early that cooperation, both within Europe and between Europe and other countries, particularly the United States, was critical to building their capabilities. However, while appreciated, the opportunities that NASA offered were not giving Europe what it really sought.

Arnold Frutkin, the director of international affairs at NASA during the agency's formative years, explained European motivation for cooperation: "An adequate outlet for European *scientific* interests may have existed in the opportunities for cooperation with NASA in the United States. But the Europeans want to master the *technology* of space activity. This accounts for the support given to the establishment of ESRO and ELDO."[4] In 1964, Europe created two organizations to pursue satellite and launch vehicle development: the European Space Research Organization (ESRO), which dealt with satellites, and the European Launcher Development Organization (ELDO), which dealt with launch vehicles. The satellite program was the more successful of the two, as ESRO launched its first satellite in 1967 on a NASA booster. All ESRO satellites were scientific in nature, which fit well with the U.S. philosophy of helping other countries without creating competition for the United States. For Europe, however, focusing exclusively on scientific missions did little to attract capital investment. Additionally, there was bitter disagreement among the ESRO member states regarding the disproportionate distribution of the industrial returns from the satellite contracts, which translated into jobs. France contributed only 19 percent of ESRO's budget, but it received 37 percent of ESRO contracts between 1965 and 1967. France argued that the disproportionality was due to its technical abilities, but France's political clout also came into play, resulting in considerable hard feelings in other countries.

Meanwhile, ELDO's straightforward purpose was to build a European rocket to break European dependence on U.S. launch technology. Europa-1, as its first effort was called, was to include parts from each participating country, and then be launched from Woomera, Australia. The individual components were of quality design, including a British Blue Streak first stage, a French Coralie second stage, a German Astris third stage, and an Italian test satellite. The combined product, however, was a Tower of Babel. By 1969, ELDO had gone 300 percent over budget and had launched nothing.

Organizational problems were as much to blame for Europa's failure as were technological issues. While cooperation was the stated goal, parochialism and protectionism narrowed what was possible to coordinate. Britain had made using the Blue Streak a precondition for program participation, but when the Europa program failed, Britain pulled out and left France with the opportunity to lead the next effort at building an independent European launch capability.

Clearly, if Europe was to compete internationally, or even not be left behind, it had to successfully cooperate internally. The limited success of ESRO and the rather dismal failure of ELDO, especially when contrasted with the spectacular success of the U.S. Apollo program, led to recognition that reorganization was needed. In July 1970, European ministerial-level meetings began to set program goals and discuss the structural reorganization needed to achieve them. Germany was interested in participating in the American shuttle program, specifically the development of Spacelab. Spacelab was a pressurized module to sit in the shuttle cargo bay and provide a pressurized "shirt sleeves" environment in which astronauts could conduct microgravity experiments, potentially leading to commercial products. The French, however, still felt that developing an independent European launcher should be Europe's number-one priority.

In 1971, the first of two key agreements, referred to as package deals, was reached by ESRO members toward establishing the European Space Agency. It transformed ESRO from an organization devoted purely to scientific research into one involved mainly in applications to satellite programs, with only a minor fraction of its budget devoted to science. Only the scientific program remained mandatory, while all application programs were optional.

The second package deal was agreed to at the European Space Conference in 1973. It focused on the à la carte program system. Arturo Russo writes:

> [E]ach ESA member state contributed to the space agency's various programs in proportion to its own political, economic, and industrial interests. Through this second package deal, France took prime responsibility for the Ariane rocket development program and agreed to contribute 63 percent of the cost. Similarly, West Germany took prime responsibility for Spacelab, a manned scientific laboratory to be carried in the Space Shuttle cargo bay, and Britain took responsibility for a maritime communications satellite.[5]

The package deal gave priority to developing both Spacelab and a new launch vehicle, Ariane (discussed in chapter 2).

Headquartered in Paris, the ESA has seventeen member states: Austria, Belgium, Denmark, Finland, France, Germany, Great Britain, Greece, Ireland, Italy, Luxembourg, the Netherlands, Norway, Portugal, Spain, Sweden, and Switzerland. Canada, Hungary, and the Czech Republic participate in some projects under cooperation agreements. The ESA has had two types of programs, mandatory and optional. The mandatory program includes basic science activities and general budget expenses. Members contribute according to their gross national product (GNP). Optional programs are those such as Ariane and Spacelab, to which members contribute according to their interests. Generally speaking, optional programs are funded by a system called *juste retour*. In theory, the distribution of ESA contracts for a program (read: jobs and industrial growth) is awarded based on financial investment in the program. If a country contributes 20 percent of the required program funding, it should receive 20 percent of the industrial contracts awarded in association with that program. In reality, however, things have not always worked that way. Initially, contracts were not awarded based on individual program investment, but on a cumulative basis. That meant that the larger countries with more developed industrial bases got the lion's share of contracts. For a while, there was even an unwritten rule that some countries could get contracts only for equipment used on the ground, because it was felt that their industries was not mature enough to be trusted to build equipment for use in space. Needless to say, such a system thwarted the entire investment rationale of the smaller contributing countries. Consequently, member states began insisting that distribution of contracts be on a program-by-program basis. That system, however, creates problems of its own, such as when a small country makes a large contribution, but does not have the industrial infrastructure or capabilities to handle an equal proportion of the often highly technical work.

The ESA focuses on R&D only. In 1976 and 1977, when Ariane was ready to enter the commercialization phase of development, some members—Germany, in particular—successfully argued that it was not the job of a body like the ESA to operate or market a launch vehicle. The bureaucratic structure of the organization would make decision making too complicated. The real reason, however, was that the ministries from the member states were willing to fund R&D, but not production and operation. This became known as the ESA's principle of "no profit, no loss." The organization engages in research, but when a technology reaches fruition, it is turned over to industrial concerns. In the case of Ariane, for example, the ESA carried out the R&D and then turned it over to a private company, though one created by the French government, Arianespace, for manufacturing, marketing, and operations.

The evolution of American–European space relations can be traced through a series of programs: Spacelab, the International Space Station, and Galileo. Each of these represents a different point in the evolution of a relationship between friends and allies, but each with different motivations, resources, and goals. Because of the technical imbalance between Europe and the United States, program cooperation has always been through asymmetric cooperative "partnerships." The United States has traditionally led, and Europe followed, including on some originally European ideas. More recently, however, Europe has decided to pursue—that is, lead—some initiatives on its own, which the United States sees as potentially competitive, even threatening.

Spacelab

As chapter 3 discusses, NASA fell from political grace after the Apollo program, and had difficulty selling the Post Apollo Program.[6] It thus encouraged European participation for very simple reasons: it felt that having international partners would strengthen the domestic political "sell" to Congress in the short run, and provide political durability in the long run. Both NASA and its political supporters recognized the value of international cooperation, dating back to early slogans of "space for all mankind" and "space for peace" that had been used to differentiate the civilian-run American space program from the militaristic Soviet program.

Europe was very interested in being involved and hoped for a partnership role, though it would clearly be a junior partnership. Initially, the European proposal was to provide some integral part of the shuttle. This was a break from the past. Previously, Europe had accepted cooperation as defined by flying experiments on U.S. spacecraft. Now, it wanted to extend beyond the scientific aspects of cooperation and into the technological. The idea of being a partner has always been important to Europe, as a partnership suggests a sense of worth in being included in decision making. Space, as this book has repeatedly pointed out, is hard to sell to politicians. It is hard for the United States, even when providing space spectaculars in return, and even harder for those who are trying to sell a supporting role. Partnership becomes important symbolically, and unfortunately for Europe, its dismal track record with the Europa launcher placed it in a poor bargaining position with the United States.

For its part, NASA felt that it was too technically risky to have Europe build a core part of the shuttle. Further, saving money was not a motivation for inviting European participation. Participation was to strengthen the program politically in the United States, and only secondarily to technically enhance the capabilities of the shuttle system through nonessential hardware. Europe's first offer of nonessential hardware was to build a reusable orbital transfer vehicle (OTV), called a space tug, to take satellites initially lifted to low earth orbit to higher orbits according to need. The Department of Defense quickly stepped in and halted that idea because of potential technology transfer issues, as building an OTV would require sharing more technical information about U.S. satellites with Europe than the United States was comfortable with. Such a vehicle has never been built, though the need for one has been consistently acknowledged.

Europe alternatively proposed building a scientific laboratory as the European contribution to the post-Apollo program, which the Germans had favored all along. Spacelab, as the proposed contribution was called, was originally conceived as part of the shuttle–space station working unit plan, though it was scrapped when the space station plans were shelved. Spacelab would allow Europe a foot in the door regarding manned space, and one that offered potential commercial opportunities down the road.

The American–European partnership on the post-Apollo program became official in 1973 with the signing of an MOU between NASA and ESRO/ESA. The MOU was laden with noncompletion contingencies for Europe, reflecting the skepticism of the United States about Europe's ability to meet its obligations. Additionally, the MOU stipulated that NASA got full control of the Spacelab hardware after delivery. In fact, Europe got one-half of the first Spacelab flight free and, after that, had to pay to use it just as any other customer would. According to some Europeans, at an estimated program cost of approximately $370 million, Spacelab was the largest gift to the United States since the Statue of Liberty.

There was some financial reciprocity in the Spacelab arrangement. The ESA provided NASA with the first Spacelab, but NASA bought an additional module as well for about $20 million. Europe had anticipated that several Spacelabs would be purchased. This was the financial "payback" expected to spur further space activity. But the modules were used considerably less often than had been projected because shuttle flights generally were much less frequent than had been optimistically projected by NASA.

European views on the Spacelab experience fell into two schools. Some felt that Europe got a bad deal from NASA; others felt that Europe negotiated a bad deal for itself to be involved in the manned space program. Perhaps most important in Europe, however, was the pervading feeling that Europe was being treated like a subcontractor rather than a partner. Many European workers, particularly in Germany, felt that Americans were condescending in their dealings with them, acting as though they were dealing with a developing country.

From the U.S. perspective, NASA was clearly being prudent in not accepting Europe's offer to build an essential component of the shuttle. The program was under significant domestic political and economic scrutiny, and relying on an unproved partner for an essential component was deemed too risky. In terms of being superior or having a condescending attitude, it should be made clear that on a one-to-one basis, scientists and engineers in the space field share uncommonly close bonds on both a personal and a professional level. They are all spurred by common goals and see themselves as pursuing a vision for all mankind. At a bureaucratic level, however, superior attitudes can prevail, not only between NASA and its European counterparts, but between companies in the United States and those in other countries. When Japanese companies were licensing launch technology from American companies, allowing the United States to "shape" Japanese launch ambitions, Japanese companies were regularly required to inform parent U.S. companies of any modifications they had made to improve performance, which they did. By their own admission, the U.S. companies rarely, if ever, even opened the envelopes that the modification notices were mailed in, let alone considered incorporating the improvements into the overall design. There was an assumption that all good ideas stemmed from the United States—not exactly the kind of attitude that shows respect for partners. For Europe, however, clearly the experience of Spacelab influenced the European attitude about joining the ISS program.

The International Space Station

In 1984, President Reagan announced in his State of the Union Address that the United States would build a space station.[7] Included in that announcement was an invitation for "friends and allies to participate in the development and use of the station." Several points about the invitation are important. First, there was not a mandate that the space station be an international program, only an invitation. Second, the invitation was for participation, not partnership. Third, this was the first of a new type of program—an open-ended program in which negotiations would involve not just hardware development, but long-term use. This third point raised both the complexity and the stakes of the subsequent negotiations to unprecedented levels. The key point for prospective participants was that they were being invited to participate in developing and using the station, but no mention was made about managing it. The implications of those distinctions quickly became apparent in negotiations for space station participation.

The station was due to begin operating in the early 1990s, but efforts fell behind schedule almost immediately, due to funding problems on the U.S. side. Political commitment to the station was weak at best. Part of the issue, as with Bush's New Vision for Space Exploration, was that once the president left the podium, presidential attention to the program disappeared.

Participation negotiations were originally to be conducted under two fundamental principles set by the United States. First, to maximize space and resources, the station would be designed according to "functional allocation," meaning that there would be no duplication of facilities for the same purpose by participants. Second, to ensure maximum use, all station elements—the individual modules or laboratories of the station, many with different capabilities—would be open to all participants. Those good ideas fell quickly to the wayside, however, victim to issues of jurisdiction and management.

Perceiving itself to have been "burned" by the United States on both Spacelab and the International Solar Polar Mission, Europe entered the negotiations with its own set of conditions for cooperation.[8] First and foremost, European countries wanted to be partners, meaning that they would have at least limited decision-making input. Certainly there was no expectation of having an equal vote, merely consultation and consideration. But because this was the first open-ended program in which cooperation inherently extended beyond development, this expectation extended to wanting a voice in the management of the station. Second, because the Europeans, particularly the Germans, were interested in potential commercial payoffs from experiments in the station's nearly zero-gravity environment, they wanted access to all the station modules. Third, Europe wanted as strong a guarantee as possible that the United States was not going to back again. Acutely aware of the inability to legally bind the United States into upholding its commitments since the cancellation of the ISPM, the European countries felt that a treaty would be their best guarantee of continued political commitment. That, in conjunction with high-level U.S. public statements reiterating its commitment, with which the United States could be embarrassed later if necessary, was recognized as about the best they could do. Clearly, the Europeans entered the space station cooperation negotiations with less than full faith in the United States as a partner.

Europe also seemed to have some unresolved issues of its own. On the one hand, Europe was clamoring for a "partnership." On the other hand, at the ESA Council meeting in January 1985, Europe had unequivocally adopted a long-term policy of developing a comprehensive autonomous space capability. Along those lines, initial proposals from Europe to NASA focused on Europe developing a free-flying or detachable version of Spacelab, with possible autonomous capability development.

However, NASA was far from keen on that idea. It suspected that Europe intended to use the space station initially, but then to eventually break away on its own. The Office of Management and Budget (OMB) agreed with NASA, but with an interesting twist that evidences one of the many internal policy debates in the United States on station design and philosophy. The OMB wanted NASA to entertain the idea of having Europe participate in building part of the core program, or the initial operating capabilities elements. That, the OBM felt, would both keep the Europeans from leaving and bring down the U.S. share of IOC costs, always a concern for accountants. But NASA rejected the idea, as it wanted to maintain full control over the core configuration. This added to the Europeans' concern about whether NASA was once again viewing them merely as subcontractors. Eventually, both sides agreed that the ESA would contribute three elements: a permanently attached pressurized module (APM), a polar platform, and a man-tended free flyer (MTFF), capable of autonomous operation for periods of six months or longer. Together, the European contribution was known as the Columbus program.

Europe's insistence on developing a spacecraft capable of autonomous operation was viewed by many Americans, and even some Europeans, as rendering European demands for a "partnership" as hypocritical. As Wulf von Kries argued,

> The Europeans have repeatedly stated that they want a "genuine partnership" with the U.S.A. Yet with their insistence on separate ownership, jurisdiction, use and control of the space station elements they intend to provide, it is quite obvious that a businesslike partnership is not on their minds. The relationship the Europeans see, in fact, seems more that of a privileged associate who, for only a limited contribution of use of his labour property, is granted a disproportionately large measure of directional rights and material benefits.[9]

Others, however, pointed out that even if Europe—or other station partners, such as Canada and Japan—had wanted to be an equal partner, it would not have been possible, because NASA wanted control. Following that line of thinking, if Europe's contribution was going to be limited and largely under NASA's control, the Europeans would have been remiss to not protect their interests as much as possible. Regrettably, but understandably, protecting national interests increasingly became the defining philosophy behind the cooperation negotiations.

When nitty-gritty station use negotiations began in the summer of 1986, NASA still supported a functional allocation approach, while the ESA did not. European concerns focused on not knowing how much it would be able to use individual modules, while being held responsible for the operating costs of some undetermined portion of the whole station. By the end of the year, functional allocation was dead. The DOD had never been in favor of the approach and had consistently encouraged NASA to drop it. However, agreement was reached that each contributor would use its own labs, but that NASA would get some use of all.

Negotiations took a sudden and unexpected turn for the worse in December 1986. Out of the blue, the DOD demanded explicit assurances that the military could conduct research on the station. This event was particularly surprising and disturbing, because in 1983, Secretary of Defense Casper Weinberger had signed a memorandum stating that the DOD had identified no national security requirements for the station. In December 1986, Weinberger sent Secretary of State George Schultz a letter, which was widely disseminated, reversing that position. While the DOD could not state exactly what it was it wanted to be able to do on the station, it clearly wanted to keep its options open. The letter stated that the United States would pay too high a price for international participation under certain conditions, including

> - Fail[ing] explicitly to reserve the right to conduct national security activities on the U.S. elements of the Space Station, without the approval or review of other nations;
> - Acced[ing] to multilateral decision-making on matter of Space Station management, utilization, or operations;
> - Permit[ting] a one-way flow of U.S. space technology to participating nations who are also our competitors in space; or
> - Allow[ing] the concept of "equal partnership" to displace either the reality or symbol of U.S. leadership in the Space Station program.[10]

As a result, in February 1987, use of the station for "national security activities" was included as part of the NASA negotiating proposal. Beyond the immediate and acute problems the letter caused between NASA and its international partners, the letter was also an early example of the growing influence of the Pentagon in civilian space matters.

During this period, the station saw its first major infrastructure reduction. In 1985, NASA had put forth a design for the station that came to be known as the "power tower," with pressurized modules clustered at one end and solar arrays for supplying power and astronomy payloads at the other. That design grew in 1986 to a grandiose dual-keel design right out of the movies. Unfortunately, after the *Challenger* accident in 1986, and remembering that Congress was less than enamored with the station, fiscal constraints soon caught up with design plans. The next year, the station was reduced to a one-keel design; NASA was also forced to adopt a "phased development" approach, in which it would build and deploy equipment according to a timetable dictated by funding availability. The ESA and the other partners were informed of these changes, but were not involved in the decisions to make them.

After national security purposes were included in NASA's negotiating principles, negotiations came perilously close to breaking down. Beyond how the DOD would figure into station activity, management decision making for developing, using, and operating the station elements presented the biggest negotiating blocks. As the primary station contributor, NASA demanded that it should have the final say on all decisions. But the ESA wanted the final say on its elements. The negotiations soured to such a degree that in November 1987, ESA delegates decided to reject the U.S. position and go ahead with the Columbus program with or without the United States, though how they would do this was not clear. Finally, a compromise was reached that gave NASA final say on the ESA's attached pressurized module, and the ESA final management decision on the MTFF and the polar platform. Furthermore, agreement was reached that the station was to be a civilian project used for "peaceful purposes"—a notably ambiguous term already discussed. In short, progress on the ISS continued by putting off the hard questions until later.

On September 29, 1988, the United States, Europe, Japan, and Canada signed an intergovernmental agreement (IGA) making them partners in the use of the space station for the next thirty years. Separate MOUs laid out the technical responsibilities of each involved agency—NASA and the ESA, for example—while the IGA was the political agreement between governments. An IGA was used rather than a treaty to avoid the potentially time-consuming Senate involvement required to sign and ratify a treaty. Though attention to the signing of the IGA was diverted by the launch of *Discovery* on the same day, marking the return of the United States to space two and a half years after the loss of *Challenger*, hopes were high that the path forward was now clear and that the space station *Freedom*, as it was then called, would soon be in orbit. That did not last long.

Between 1988 and 1991, funding for *Freedom* was repeatedly cut, forcing more design downsizing. As part of the downsizing, power supplies were cut dramatically. Power to operate centrifuges requisite for microgravity research of potential interest to commercial markets—a key motivation for European, particularly German, participation—was disappearing. Japan's official reason for participation in the space station was to conduct microgravity research as well. It, too, was left scrambling. Ironically, the "real" reason for Japanese participation was to demonstrate to the United States its reliability as a partner on a large, complex, open-ended space program. The reliability of Japan has never come into question, though that of the United States certainly has.

At the end of 1990, the *Freedom* design collapsed again. Reviewers found that the program was over budget and that the design was too heavy and complicated (requiring too many spacewalks by astronauts for both assembly and maintenance) and provided too little power. In March 1991, NASA unveiled its new space station, mockingly referred to by critics as "Space Station Fred." The partners, again not included in the redesign, were understandably dismayed.

Politically, the collapse of the Soviet Union then began to play into the future of the space station. Between 1991 and 1993, European consternation, American inability to come up with a viable station design, Russian hopes to keep its space legacy alive, and American desire to keep Russian rocket scientists gainfully employed created some interesting scenarios. In 1992, Russia and the ESA discussed collaborating on the development and use of a Mir-2 station. That did not sit well with the United States, which initiated talks of its own with Russia. Subsequently, in November 1993, NASA unilaterally announced Russian inclusion in the space station program, with incorporation of the European and Japanese modules, to develop a single space station, the ISS. Yet again, it was a whole new day for the original partners.

The 1990s were not easy years for the ISS. Including Russia and redesigning to its strengths created both opportunities and pitfalls. Politically motivated and prodded by the Clinton administration, NASA administrator Dan Goldin liked the idea of expanding the partnership, to the extent that he jettisoned the rule that only NASA could build IOC hardware. Goldin believed that by turning over responsibility for some of the station's most critical components to Russia, including the lifeboat, unmanned cargo carrier, and first habitable module, NASA could save the cost of developing them, and facility construction could be accelerated. Cash-strapped, Russia squeezed every dime it could from the United States in negotiations for taking over those responsibilities, and often fell behind on deadlines, but for the most part eventually met its commitments. Waiting on the Russians and with funding always in jeopardy, NASA was constantly rewriting ISS time lines. The redesign after the Russians joined the program did not finalize the blueprints either, as all the parties dropped modules and capabilities. Nevertheless, the Russian-built Zarya control module was finally launched in 1998 as the first module for a station program first announced in 1984. Between 1998 and 2004, assembly continued, and on-board work commenced when the first "occupying" crew arrived in 2000. Meanwhile, Europe and Japan wait. If all goes well, the European Columbus module, the largest piece of hardware remaining from the original European plan, should be launched in 2007. The Japanese module is scheduled for 2009. Those dates, however, are highly questionable, because of ongoing problems with the shuttle and NASA's shift in emphasis toward moving ahead with the Bush's vision for returning to the moon and traveling on to Mars.

In 2004, Russia announced that after 2005, in accordance with the original interagency agreement signed in 1998, it would cease transporting American astronauts to the ISS for free and would provide access only on a commercial basis. The Russian space program is relatively healthy again, but its budget is still tight, and Russia has learned the lessons of capitalism well. An agreement is an agreement, even among partners. Additionally, the last Soyuz transport, which serves as the emergency rescue craft for the station, is scheduled to leave the station in 2006.

The easy answer might seem to be for the United States to simply "buy" Russian launch services, potentially even to finish the ISS. But funding for completing the station, already in competition with the Bush vision, is rigidly tight. There is no slush fund for unexpected expenses. Further, buying Russian launch services does not relieve NASA of the need to keep the shuttle flying, as some pieces are simply too big to be transported to space any other way.

Also, negotiations for Russia supplying additional spacecraft were complicated by politics. According to the Iran Nonproliferation Act of 2000, NASA cannot buy anything from Russia unless the president first certifies that Russia is not sending nuclear technology to Iran, which it cannot. In July 2005, Secretary of State Condoleezza Rice and NASA administrator Mike Griffin sent Congress a letter saying that the administration would soon ask to amend the Iran Nonproliferation Act to allow the purchase of Russian flights. That request was subsequently made and granted in October 2005, evidencing the dire state of the U.S. manned space program. In January 2006, NASA announced that it would pay the Russian company Roskosmos $43.8 million to transport one astronaut to and another one back from the ISS—another large sum of money unavailable to either fly the shuttle or build the crew exploration vehicle.

Without the shuttle, the United States and other partners rely on Russia for transportation to and from the station. The shuttle is due to be retired as soon as the ISS is completed in 2010. But the ISS is scheduled to operate for five years beyond completion, and the United States may have no way to get there. If all goes perfectly, which rarely occurs in matters concerning space, the proposed CEV will not be flying until 2014. So under the best of circumstances, a spacecraft that took twenty-six years to build will be used for only five years after completion. Even worse, after having spent almost $100 billion in taxpayer money, the United States could be left unable to send crews to the completed station, leaving the Russians in control and potentially making money hand over fist selling tourist tickets for a visit to space. Aerospace engineer and investment manager Dennis Tito paid $20 million to the Russians for his visit to the ISS in 2001, and others have followed since.

The partners, however, are not as anxious to abandon the station. Europe and Japan have yet to even put their modules into orbit. Once the modules are in orbit, science programs that have long been structured around their use are ready for execution. At a January 2005 meeting in Montreal of the heads of the space agencies of the United States, Russia, Europe, Canada, and Japan, the partners avoided any discussion of the new U.S. initiative. Their priority is to complete the ISS. At a meeting between the deputies, the Russians objected to the wording of the draft final meeting report; they felt that its ambiguities left the door open for the United States to potentially withdraw from the ISS program, which all the partners still fear will occur. Dependence on the United States for space capabilities has proved to be a shaky proposition for the ISS partners. Dependence on Russia for transportation to the ISS has left the United States in a vulnerable position as well. It is understandable that dependence is not a strategic goal that countries often set for themselves; even among friends and allies, it requires trust.

Meanwhile, Russia is pursuing the development of a reusable *Clipper* shuttle. If successful, and the Russians hope there will be test flights early in the next decade, that vehicle could carry a crew of six to the ISS, three more than current Soyuz transports. The ESA is very interested in partnering on this Russian program. While funding held up its commitment to the program in 2005, ESA director-general Jean-Jacques Dordain remained confident it would eventually be approved.

Even beyond space activity specifically, developed countries increasingly view developing and safeguarding certain space capabilities as necessary to maintain their own sovereignty. Financial transactions, transportation, domestic and international communications, the ability to monitor and potentially anticipate environmental disasters, and conflict management—all relevant to a country's sovereign ability to manage its own security—rely more and more on space hardware. Regardless of assurances given by the United States that it will monitor and maintain the hardware for the world, for some countries, experience and prudence dictate otherwise. And the United States often views the space activities of others, even friends, with chagrin.

A Mature European Program

The first Gulf War in 1990 and 1991 was considered the first space war because it was then that the value of space assets as force enhancers was first recognized. While many coalition countries saw the value of space assets, the ability to reap their benefits was largely limited to the United States and those to which the United States parceled out the benefits. Coalition members largely depended on the United States for remotely sensed imagery, and many allies felt that the United States was often stingy.

After the Gulf War, several countries, including some that had defined the "peaceful" use of space as meaning nonmilitary, changed their view and began developing dual-use space technology, such as imagery, for uses including those for the military. Great Britain has enjoyed special access to U.S. imagery and has been less interested than other countries in developing autonomous capabilities. Germany, however, is building a series of radar observation satellites. France launched its third military spy satellite, *Helios 2A*, into orbit in December 2004. It is said to be able to spot textbook-size objects anywhere on earth and is equipped with infrared sensors, allowing it to gather information at night as well as during the day. French president Jacques Chirac argued in 2002 that without its own satellite capabilities, Europe would remain little more than a "vassal" of the United States.[11]

The United States has not encouraged Europe to develop military space hardware. The rationale has been that European military needs are great in so many other areas, such as transport planes and precision guided munitions, that it should prioritize its spending there rather than developing spy satellites. The U.S. view has also been that there is no need, because high-resolution imagery is now commercially available. When ordering imagery from a U.S.-affiliated company, however, countries reveal what they are interested in looking at, and that information can find its way to the U.S. government. Further, many countries simply do not trust the United States not to exercise shutter control and switch off commercial imagery in a crisis.[12] The European prioritization for development of its Global Monitoring and Environmental Security (GMES) system, an autonomous European dual-use satellite monitoring capability, reflects this distrust. GMES, Galileo, and Ariane represent the three pillars of Europe's current space strategy, as they all provide autonomous capabilities.

Europe is not alone in its desire for autonomous capabilities. Japan, too, has changed its attitude about dependence. Japanese politicians were unhappy with their access to information and imagery from the United States after the North Korean Taepo Dong missile flew over the Sea of Japan in 1988.[13] Subsequently, Japan has invested over $2 billion in a dual-use satellite system called the Information Gathering Satellite (IGS) system, which provides imagery no better than that available commercially, but which it controls so nobody else knows what Japan is observing. Autonomy was its key motivation.

A November 2003 EU white paper posited autonomous capabilities in space as critical in areas ranging from environmental protection to internal security. That view is further reflected in the December 2003 EU report "Space and Security Policy in Europe," which states, "Space is a strategic asset, and its importance both in terms of technology and security cannot be overstated."[14] Security, however, can be defined in many ways. The United States defines it primarily in militaristic terms, but Europe and others do not: "Space technology is linked to collective security, with the term 'security' referring to the protection of European citizens from potential risks of both military and non-military origin."[15] This broad definition recognizes the human security needs of freedom from pervasive threats to people's rights, safety, and lives. It includes items such as clean air, clean water, food, the potential for economic growth, and physical protection.

Regarding economic growth, in the globalized economy, developed nations cannot rely on the routine production jobs of the past to maintain a middle class. Those jobs are in many cases moving to developing countries with lower wages. Service-sector jobs have replaced production jobs in many instances. While service sectors in the past have traditionally offered low wages, new service areas—such as home security, computer geeks who make home visits, and urban dog-walking and boutique services, such as $200 haircuts—have changed that somewhat, but not enough to support an entire middle class. More important is the notion that in the future, strong economies will be knowledge-based. Europe has stated that it will develop a knowledge-based society by 2010.[16] In knowledge-based societies, worker input is measured in value added by knowledge, rather than manual labor. Because of the linkages among space, information technology, and sharing and transmitting knowledge and its work products, countries seeking to keep up with the new millennial capitalism deem access to space technology as essential. The South Korean government, trying to get ahead of the knowledge curve, has instituted a policy toward providing broadband Internet access to all its citizens.

Europe also hopes that developing GMES and Galileo will slow its brain drain. Almost 70 percent of the Europeans who receive doctoral degrees in the United States in science and engineering opt to stay in the United States. Reversing that trend would be of considerable long-term value to Europe in building its knowledge-based society. Galileo alone is estimated to create more than 100,000 European jobs,[17] and will provide highly trained individuals with jobs in their fields. European Union officials hope that Galileo will be the first of many such programs.

Additionally, space technology is viewed as critical to develop and execute integrated policies domestically, regionally, and globally. Coordinating airline traffic, or even rail or automobile traffic, is significantly enhanced by navigation satellites and communication systems. Space-based earth observation systems can assist in crop rotation to increase yields and help identify areas of land with high groundwater contact for spraying against malaria-carrying mosquitoes. Satellites carry news and weather broadcasts to remote areas. The number of ways that space-based technology can and is being used in civilians' everyday lives is increasing exponentially. But Europe wants space technology for more traditional security reasons as well.

European Military Space Efforts

Since the European Defense Community proposal failed in the 1950s, Europe has been trying to develop a vehicle for a European concept of a common defense, independent from but closely coordinated with NATO. In 1992, representatives of twelve European countries signed the Treaty of the European Union, also known as the Maastricht Treaty, paving the way for the current European Union. The treaty was designed around three pillars: economic and social cooperation, a common foreign and security policy (CFSP), and police and judicial cooperation in criminal matters. While great strides have been made in the first area, events in the former Yugoslavia showed that, unfortunately, the European states still have a very weak political will and capacity to carry out their own policies, leaving it to the United States and NATO to take the lead.

The development of an independent European security and defense policy (ESDP) was launched by the European Council in June 1999 as a distinctive part of the CFSP. The stated intent of the ESDP is to strengthen the European Union's external ability to act through developing civilian and military capabilities for international conflict prevention and crisis management. The reaction of the United States to both the CFSP and the ESDP has ranged from ambivalence to enmity, welcoming the idea that Europeans could do more in their own defense, but concerned that their efforts could undermine transatlantic ties and NATO. Both the CFSP and the ESDP require many new military capabilities; both seem also to point to the need for a European space security policy. Such an integrated policy approach has been considered since at least 2003, but is still emerging along with the ESDP and the CFSP. A major stumbling block is that, to date, most European military space programs have been strictly national programs.

Earth observation is the only field in which there has been prior tangible cooperation between European governments. Since 2001, European governments have used the EU satellite center in Torrejón, Spain, for imagery interpretation, including some imagery from the French Helios military satellite system. Since 1999, Belgium, France, Germany, Italy, and Spain have been working to establish common requirements for future observation satellites. While Belgium, Italy, and Spain purchased shares in Helios so they can use it, the satellite was purely of French design. If common requirements can eventually be agreed on, the development of joint spy satellites might be in Europe's future.

Beyond earth observation, efforts have also been made to integrate telecommunication satellite networks. A NATO satellite has been under discussion, but no agreement has been reached because the United States is involved. European governments have resisted, as their requirements are not nearly as large as those of the United States. They do not want to buy a Lexus (which the United States usually does) when they need a Chevy. Other potential intra-European efforts in the future may be in the areas of early warning, electronic intelligence, and space surveillance. The political and economic obstacles, however, mean that efforts will not proceed quickly.

Part of the problem is that Europe must decide exactly what it wants to do and how to pay for it. The U.S. approach to space has been broad, increasingly expanding to include space control and potentially force projection. Europe's military space ambitions are narrower. Space control technology is mostly irrelevant for the kind of conflict prevention and crisis management missions Europe envisions. European missile defense concerns are also much narrower, focusing on short- and medium-range missiles rather than the long-range missiles of greatest concern to the United States. Observation and telecommunication are first on European priority lists for their intended missions. Clearly, the views are spurred by different strategic perspectives. The United States maintains a truly global strategic outlook. Europe, however, has far more limited ambitions and, hence, requirements for force projection, which are focused on relatively close threats. Such a regional focus does not exclude the possibility of force commitments on a broader range, but they would clearly be in support of, most likely, American efforts, rather than Europe taking the lead or initiating global military action on its own.

The differing goals of the United States and Europe toward space are also reflected in their comparative spending. As discussed in chapter 3, estimating U.S. expenditures on military space programs is very difficult. Nevertheless, it is estimated that U.S. spending accounts for approximately 90 percent of the world's spending on military space programs. The unclassified portion of the DOD budget request for space was $22.5 billion for fiscal year 2006, compared with $19.5 billion in 2005.[18] Europe spends, in total, about 5 billion euros ($6.2 billion) annually on space, which covers civilian and military, national and cooperative efforts.[19] While European space spending is expected to grow as a percentage of its gross domestic product (GDP),[20] Europe clearly has neither the desire nor the resources to attempt parity with the United States.

In developing future capabilities, the United States sees the dual-use nature of space technology as problematic, but Europe regards it as an opportunity. Space assets are viewed as a means to protect populations, resources, and territories, as well as maintain the integrity and capabilities of the technological base. Dual-use national programs are already in existence or planned. In 2001, France and Italy agreed to develop Pleiades-Cosmos as a civilian satellite observation system, while acknowledging that it could also be used for military purposes. Being able to use these high-cost assets for not one mission, but potentially two, makes political justification substantially easier:

> The development of dual-use technologies calls for a "European" approach to space security, linking the present national defence programs with mainly civilian European programs. The functions and means of security and defence uses of space overlap considerably. In fact, space operations can be seen as a continuum, including civilian and military functions as well as security and defence operations.[21]

Clearly, European attitudes about dual-use technology diverge considerably from those of the United States, which, as chapter 2 discusses, seeks only control.

Europe's GMES and Galileo are its most ambitious autonomous space initiatives to date, and both have military implications. Both are also exceptions to the rule, in European countries' ability to jointly develop space programs of that magnitude. Previously, Europe has staggered, plagued with funding problems and political and institutional fragmentation when it came to implementing large space programs with potential military use. That Europe politically came together to support the programs, largely to break reliance on the United States and build its own industrial capability, reflects the programs' perceived importance.

Within Europe, moving from traditional space missions to those of a more security-oriented nature initially created somewhat of a turf war. The ESA had long held the reins on joint European space activity. But the ESA was and is basically a counterpart of NASA. Its interests are guided by those whose priorities are focused on space, not security. The European Union, however, recognized the value of space technologies for a variety of purposes, including environmental protection, but also extending to internal security and managing autonomous military operations.

Kludging EU and ESA interests initially looked impossible. After consideration, however, European authorities decided that the ESA's civilian orientation does not preclude military operations, nor are such operations legally prohibited. The preamble of the European Space Agency Convention makes it clear that ESA missions will be for "peaceful purposes." But using space to "help secure peace and defend stability" is considered to be compatible with the nonaggressive use of dual-use technology. Even new institutional parameters are being considered:

> The institutional separation of civil and military space activities was historically rooted (as with NASA and the US Department of Defence) and was originally based on valid political and legal considerations. However, it became increasingly outdated after the Cold War. . . . There is the potentially attractive option to take full advantage of the dual-use nature of space in ESA itself, based on a future co-operative agreement with the EU. Any such opportunity to avoid intra-European duplication should be welcome as a cost-reducing factor.[22]

Europe is rethinking its organizational structure to maximize scarce resources. It appears that the ESA and the EU will share power on space issues, with the EU organizing policy and the ESA implementing programs. The European Union has much to gain in that regard.

As with European development of Airbus and Ariane, Galileo shows the European Union encouraging innovation, as well as creating high-quality, knowledge-based jobs. Furthermore, high-utility projects, such as Airbus and Galileo, make the benefits of the EU more visible to European businesses and citizens. In other words, the success of Galileo will add credibility to the European Union as an institution, potentially contributing to its long-term success and expanded responsibilities. However, there are limits to the willingness of European countries to give up their sovereignty, even if potentially for the greater good, as the failure to ratify the Maastricht Treaty in France and the Netherlands in May 2005 demonstrated.

While Galileo provides Europe a seat at the security table, it is important to point out that Europe does not view Galileo's strategic value in the same sense as it views, say, nuclear weapons. From a techno-nationalist, geostrategic perspective, it indicates power. It does not and does not intend to put Europe into competition with the United States as a global military power, though it impinges on an area the United States considers strategically important and had monopolized.

Galileo

Access to the GPS system operated by the United States is free worldwide. Civilian use has made it a global utility. Yet other countries are acutely aware that the U.S. military controls the system and could deny it to others if it chose.

The chances that the United States would actually turn off the GPS are slim at best, as the United States relies more on its capabilities for civilian and military purposes than does any other country. But the possibility of denial, and relying on the largesse of the United States, has contributed to a growing number of countries seeking their own space capabilities, such as navigation satellites, including those with military potential.

After initial discussion in 1999, in 2002, European transport ministers decided to support Galileo. The thirty-satellite system is scheduled to begin operations in 2008; the first satellite demonstrator, *Giove A*, was launched on December 28, 2005. Program development will cost approximately $4 billion and is a publicly and privately funded venture, with public funding coming from both the transport and defense ministries. Interestingly, not everyone in the defense ministries initially supported Galileo:

> At the height of the European debate on whether or not to fund the programme in January 2002, the then French defence minister, Alain Richard, indicated that "he saw no compelling military case for Europe to launch its own fleet of satellites to match the GPS network already in space." Afterwards, an unambiguously supportive statement from the French foreign ministry contradicted Richard's position. But Richard's views represent a traditional conservative position, often found in European defense ministries, that prefers the status quo. Moreover, very tight national defence budgets limit European choices. Spending money on space programmes has never ranked high on most defense ministries' priority list, except for strictly controlled national programmes like the French Helios series.[23]

It has taken longer for the hang-together-or-hang-separately approach that has strengthened European abilities in the civilian space arena to take hold in defense.

Washington has several concerns with Galileo. First and foremost, there is simply the issue of losing control. Any increase in capabilities by another country is viewed as a relative decrease in capabilities because of the American zero-sum attitude toward space. If Europe (or any country) gains, the United States must lose. Space assets are so important to the U.S. military that space dominance is deemed critical, and any increase in capabilities by others is seen as diminishing that ability to dominate. There are more specific economic, technology, and military concerns as well.

Initially, the United States was concerned about European plans to overlap its commercial radio frequency signal with that of the U.S. military's classified signal. Meetings on that topic, over a four-year period, were heated on both sides. The United States wanted to be able to jam Galileo signals without affecting its own GPS military signals. An agreement reached in November 2003 between Europe and the United States addressed many of both parties' concerns: Europe agreed to modify Galileo's signals, and the United States agreed to give Europe technical assistance in developing Galileo to make sure that the European system and the third generation of GPS, to be deployed in 2012, will be compatible. Compatibility is important to ensure the interoperability of the two systems, a commercial goal of both sides. However, it could also give Europe the ability to jam U.S. signals. That possibility might be where European concerns about counterspace operations, including blowing satellites out of the sky, come into play.

Besides commercial operability (discussed in chapter 1), the United States has been very concerned about international cooperation on the Galileo program. Russia is interested in creating synergies between Galileo and the Russian Glonass navigation system. Both China and Israel signed partnership agreements with Europe to become shareholders in the Galileo system; both countries made cash deposits with the Galileo Joint Undertaking office in Brussels to guarantee their inclusion. The concern of the United States focuses primarily on the October 2003 agreement signed with China.[24]

There is no evidence that any country has yet used GPS-guided equipment against U.S. forces in combat, but that could change. Iraq tried to jam GPS signals in 2003, but that turned out to be a big mistake for the Iraqi jammers, as they brought on their own demise when the Air Force was able to lock onto their signals and subsequently rain missiles on their location. Pentagon officials worry that in the event of a conflict over Taiwan, China could use GPS to target weapons against the United States.

However, the agreement between Europe and China on Galileo has been largely an empty shell. While China invested some 200 million euros in Galileo, most of that sum was spent in China to build ground installations and to promote domestic use. Beyond that, China wanted to use its Long March rockets to launch Galileo, but the United States objected to that based on potential technology transfer issues regarding the satellites themselves. Since certain ESA programs, including Galileo, include critical components, the United States can veto such proposals: dependence rearing its head again, with accompanying fears of retaliation should Europe defy the United States. According to a 2004 document of the Western European Union: "If Europe pursues cooperation with China on the Galileo programme, we run the risk of encountering opposition from the United States, which could mean having to envisage the possibility of producing critical components ourselves, unless a way can be found of reconciling everyone's interests before the United States decides to adopt retaliatory measures in other areas."[25]

Issues created by China as a stakeholder in Galileo could, however, soon become moot. The Galileo Joint Undertaking office will be replaced by a new Europe-only management company, the Galileo Supervisory Authority. Consequently, it is expected that the shareholding partnerships with Israel and China will expire at the end of 2006 when the management change occurs, and their cash deposits are likely to be refunded. Not surprisingly, China has been very upset by this change of circumstances. It has countered with its own plans, or threats, to build its own navigation satellite system, called Compass, and it has apparently started purchasing the requisite atomic clocks for the system.

Difficulties with the Sino-European Galileo agreement notwithstanding, there is a broader plan of cooperation between Europe and China, and it has a significant history. For twenty years, China and Europe have worked together on space-based earth observation programs and on launching observation satellites. More recently, China and Europe have been working together on China's Double Star program, its first mission to explore the magnetic field surrounding the earth. Indicative of its name, the mission consists of two satellites in complementary orbits, designed to simultaneously gather data on the changing magnetic field. China designed, built, launched, and operates the satellites. The intent is to have China's satellites work in concert with four ESA satellites, together called the Cluster mission. The ESA satellites were launched in the summer of 2000 into elliptical earth orbits; the Chinese satellites in 2003 and 2004.[26]

Commercially, the French satellite manufacturer Alcatel, which ranks third in the world and first in Europe, has had a presence in China since 1983 and earns 10 percent of its income there.[27] In 2002, Alcatel and the China Aerospace Science and Technology Corporation (CASC) signed a contract for the joint development of the first Chinese high-capacity communication satellite. Snagging that contract was a feather in Alcatel's cap, and Alcatel intends to double its business in the region over the next several years. Remember that following the Cox Report, U.S. satellite manufacturers have been unable to sell communication satellites to China and that services are an even more lucrative aspect of space-related commercial sales than is hardware. That China has opted for the European standard for its mobile-telephone technology is important for future service contracts. Clearly, Europe has made significant inroads into the potentially lucrative Chinese market and intends to expand them.

More broadly, "the EU has expressed the intention of developing its strategic partnership with that country,"[28] meaning China. Europe's willingness in 2005 to consider lifting the arms embargo on China, which has been in place since the Tiananmen Square massacre in 1989, evidences the extent of that intention. In response to even the suggestion of lifting the embargo, Congress passed a resolution declaring that such an act would be inconsistent with transatlantic defense cooperation, and threatened constraints on the transatlantic defense relationship. Not surprisingly, France pushed lifting the embargo, seeing clear economic advantages and the potential to create counterbalancing pressures on unilateral U.S. power. Because the United States views China as a potential peer competitor, if not an enemy, it saw Europe's action as a betrayal. Europe recognizes that a potential strategic partnership with China is not without limitations and pitfalls, but sees the potential benefits as worthwhile. The United States has seen only pitfalls.

Talk of lifting the embargo stopped when China passed an antisecession law in March 2005, threatening use of "non-peaceful means" if Taiwan crossed the line toward independence. The European Union subsequently decided not to reconsider lifting the embargo until 2006 at the earliest—avoiding a confrontation with the United States as well. Apparently unable to resist reverting to its bullying manner of the past in this case, China paid a high price for codifying its long-standing threat to use force.

A multitude of unresolved issues remain regarding increased Sino-European space cooperation. Observation satellites are considered to be warfare equipment by France's Interministerial Committee on the export of war equipment. While not necessarily banned for export, that potential makes them more complicated. It is generally considered, however, that if China is seeking to buy metric (non-U.S.) observation hardware, it will be able to get it, and if not from Europe, then from Russia, Israel, or elsewhere. Unlike the United States, Europe is reluctant to try to control what it does not monopolize.

Europe has its own issues with China and has stated that Sino-European space cooperation would be greatly facilitated if China would make "good faith" efforts in several areas. First, like everyone else, Europe would like China to be more transparent and forthright about its space programs, in terms of both activities and intentions. Second, Europe wants China to join the Missile Technology Control Regime, a structure intended to stop the proliferation of missile technology. Third, Europe has encouraged China to make greater efforts to reach an understanding with the United States regarding potential Chinese inclusion on the ISS, though the United States has resisted it.[29] There have also been calls for China to set up a civilian space structure equivalent to that of Europe, which is rather ironic given European discussions regarding the inefficiencies of Cold War structures that separate civilian and military space activities. Finally, Europe has pushed for Chinese ratification of the United Nations Covenant on Civil and Political Rights, because China's human rights situation is linked to potential European willingness to lift the post-Tiananmen arms embargo.[30]

Economics and competition for technologies that spur economic growth have consistently motivated space activity in Europe. That the United States was just as interested in the commercial ramifications of making Galileo and the GPS compatible is evidence of the global nature of that interest. The dual-use nature of space technology, however, adds military considerations to what otherwise might be commercial questions. And, unfortunately, though economics is not a zero-sum game, military security issues can be viewed that way and are by the United States. Therefore, the question is whether the United States and Europe can cooperate in some areas as they compete in others, and whether Europe can cooperate with others and the United States at the same time.

Clearly, Europe's excursion into space with the Galileo program has attracted the attention of the United States. Ignoring Europe and the Galileo program has not been an option. That certainly confirms the strategic nature of space, given American willingness to discount European views and activities in other instances. From the European perspective, however, space is strategic in very different ways from those prioritized by the United States. The GPS is a global utility owned by the United States, but by definition, utilities are essential. Countries will feel compelled to develop their own utilities to protect their sovereign choices, and Galileo's potential for helping to build a knowledge-based society in Europe is another strategic benefit.

Europe's desire to work with the United States while maintaining its autonomy and the option to work with other countries as well will create problems with the United States, especially if Europe chooses to do business with countries that the United States perceives as potential threats. The alternative, however, from Europe's perspective, is to put its fate in the hands of the United States. Partnerships with the United States are inherently difficult because they can never be true partnerships; they are economically impossible for Europe, and politically unacceptable to the United States. For the United States to expect Europe not to be interested in exploring other options is unrealistic, especially when history has shown that American–European cooperation has not always met European expectations.

Europe has neither the resources nor the ambition to seek parity with the United States in space. It needs a combination of autonomous capability development, as well as cooperative programs with the United States and others. The United States may see this agenda as a betrayal and try to stifle European plans. However, while stifling European plans might be potentially effective on individual programs or in the short term, it could also provide further impetus for more autonomy and alternative partnerships. Increasingly those alternative partnerships include or even prioritize China and developing markets seeking a greater presence in space, where Europe can lead or co-partner the relationship rather than be a junior partner at best, as is the case when working with the United States.

The Ambitions of China

EIGHT

> The lunar exploration project will have an incalculably valuable effect on the ethnic spirit and motivation [of the Chinese people] and I ask you, how much is that worth?
>
> —Ouyang Ziyuan

> For countries that can never win a war with the United States by using the methods of tanks and planes, attacking an American space system may be an irresistible and most tempting choice.
>
> —Wang Hucheng

In November 2000, the Information Office of the State Council issued the first Chinese white paper on space: "China's Space Activities."[1] The technical milestones it laid down were impressive, and the language was assertive. It reminded readers that China invented gunpowder, the "embryo of modern space rockets." China wants to regain a place of distinction in a field it sees itself as having initiated and once dominated. In that way, China's space ambitions are unique. In others, however, they represent the high end of ambitions of many developing countries. The latest Chinese white paper on space was issued in October 2006.[2] It provides considerable detail about the aims, principles, and accomplishments of China's space program and recognizes that "the role of space activities in a country's overall development strategy is becoming increasingly salient." China states its support for international cooperation:

> [the] Chinese government holds that outer space is the common wealth of all mankind and each and every country in the world enjoys equal rights to freely explore, develop and utilize outer space and celestial bodies; and that all countries' outer space activities should be beneficial to the economic development, social progress of nations, to security, subsistence and development of mankind, and to friendly cooperation between people of different countries.

While emphasizing that "China will focus on certain areas while ignoring less-important ones," satellites, launch vehicles, launch sites, telemetry, tracking and command, manned spaceflight, and deep-space exploration are cited as areas of interest and achievement.

Realistically, some of these goals are out of reach for many developing countries, at least in the near term. But for many, they are nevertheless inspirational. Brazil has ambitious space plans, but insufficient funds to support them. In its third attempt to launch satellites into space using its first family of Veiculo Lancador de Satelites (VLS) rockets on August 22, 2003, the rocket exploded, killing twenty-one engineers and technicians. A successful launch would have made Brazil the first Latin American nation to send its own satellites into orbit. Subsequently, in March 2006, a Russian Soyuz capsule ferried not just a Russian and an American astronaut to the International Space Station, but Brazil's first man in space, Marcos Pontes. During a visit to Brazil in November 2004, President Vladimir Putin pledged that Russia would assist Brazil in rebuilding its space program and restoring its launch base, clearly seeing cooperation as advantageous to both countries. President Hu Jintao of China also visited Brazil's space center that month, as Brazil and China have already cooperated on two satellites.

When countries like Nigeria and Venezuela contract for their own satellites while a surplus of commercial transponder capabilities exists, the techno-nationalist proclivities of space activity are clear. An Iranian microsatellite was launched into space on a Russian rocket in 2004, raising concerns about Iran's future plans. For many countries, the idea of moving from being a space user to having independent capabilities has security dimensions. India has access to its own remotely sensed data, while regional adversary Pakistan is largely dependent on what it can buy commercially, potentially giving India a strategic advantage.

In an article and book, Thomas P. M. Barnett suggests that "connectivity" is a key consideration for attaining the desirable and far safer status of being a "core" country rather than a weak and dangerous "gap" country.[3] Space technology plays a big role in connectivity and, hence, is a tool of globalization—one that might keep a country from being left behind economically and prevent a visit from the U.S. military. In the article, Barnett says, "It's always possible to fall off the bandwagon called globalization and when you do, bloodshed will follow. If you're lucky, so too will American troops." The idea is that if a country is not succeeding in globalization, internal strife is likely part of the reason. However, not all countries see the prospect of American troops arriving as a good thing, regardless of intentions. Successfully globalizing is a key goal for developing countries, and space offers undeniable benefits, from disaster mitigation and management to urban planning, maximizing crop yields, and connectivity.

In some ways, China clearly is not the average developing country (whatever that is). China has a population of more than 1.3 billion people and is viewed by much of the world in terms of market potential, as its economy has been growing by double-digit rates. China's resources and political will are substantially greater than those of many developing countries. While a considerable amount of Chinese resources go to keeping the country from imploding, government resources for priority projects are available at a higher level than they are in many countries, especially related to security. Ample cheap labor is available as well. Having a centralized government also allows for top–down decisions to support areas that might otherwise not garner political support, such as a manned space program. That China has a manned space program at all makes it atypical among developing countries.

Finally, China is a rising power. The implications of that status are debated in political circles almost endlessly. Analysts, including Henry Kissinger, Benjamin Schwartz, and David Shambaugh, have said in one way or another that simply trying to contain China would not work, and that "managing" China's rise, as Schwartz puts it, would be the more effective strategy.[4] Economics alone dictates that the United States pay attention to China. Considering both the growth rate of the Chinese economy and the market potential still untapped, the United States must consider the possibility that there could be a time when simply outspending China on technology will not be a viable strategy for staying ahead militarily. But the Blue Team—those who vocally and voraciously viewed China as the next enemy of the United States during the Cox Committee hearings—does not view "managing" China as an option, and many of its members and supporters have gone from being Washington outsiders during the Clinton years to being insiders with the Bush administration. Their views regarding China largely remain belligerent: their goal is to stop the rise of China, and they equate engagement with appeasement.

Regionally, China has done remarkably well since the 1990s in transforming its regional image from that of a "bully" to a "power," and the image transformation extends beyond Asia as well. A Pew Research Center poll taken in April and May 2005 is indicative. According to that survey, "strikingly, China now has a better image than the U.S. in most European nations surveyed."[5] In terms of space, this positive image means that many countries view Chinese activities as creating opportunities for potential partnerships. By contrast, for conservative and Pentagon analysts who see China as America's most likely next peer competitor, Chinese space activity is regarded as creating more challenges.

Communication between Chinese and American government agencies has been limited and formal. Interactions between NASA and the China

National Space Agency (CNSA) have been rare: invitations from NASA to CNSA are seldom offered, and visas for Chinese participants are often denied if space events in the United States are open to the public.[6] Official U.S. participation in Chinese-sponsored space workshops or events has been a near impossibility, as the United States has not wanted to signal intentions that it is not prepared to support.

Though rare, there have been notable U.S.–China meetings on space. In November 2004, a Chinese delegation was invited to attend a three-day workshop in Houston on President Bush's moon–Mars initiative. Chinese attendance at these kinds of events requires the blessing of the Department of State, so the workshop was considered somewhat of a break from the past. Additionally, former NASA administrator Sean O'Keefe welcomed CNSA administrator Sun Laiyan to NASA headquarters in Washington for a courtesy visit—translated as "no business was discussed"—on December 2, 2004. From 2004 to 2006, however, contact was sparse. Then, in April 2006, CNSA vice administrator Luo Ge headed a delegation visiting NASA's Goddard Space Flight Center and gave a symposium presentation in Washington. There, he reiterated a year-long invitation to NASA administrator Mike Griffin to visit China. Though Luo also said that China had not been invited to discuss specific lunar cooperation projects as part of NASA's new effort to explore the moon, it had been invited to attend the 2004 workshop, and a delegation from China was again invited to the NASA "exploration strategy" later in April 2006.

In January 2006, a three-person bipartisan congressional delegation visited the Jiquan launch site, used for manned launches, the first U.S. delegation to be allowed to visit. Representatives Mark Kirk (R-Ill.), Rick Larson (D-Wash.), and Tom Feeney (R-Fla.) are members of the congressional U.S.-China Working Group. News coverage of the visit was fairly widespread in Asia, but it went virtually unnoticed in the United States. The American press did report that the delegation was carrying a mandate from President Bush to look for areas of space cooperation that could be part of the agenda between Bush and Hu during Hu's upcoming visit to Washington, which would have indicated a major administration policy change.[7] The existence of such a mandate was later refuted by the group, however. Potential cooperation was discussed only in developing mutual rescue capabilities. Interaction between the United States and China on military space issues is even more rare. While Air Force general Richard Myers, then-chairman of the Joint Chiefs of Staff, led the first American delegation to the Chinese space center outside Beijing in 2004, the Chinese facilities tour was restricted at best, and did little to counter U.S. frustration about the opacity of the commingled Chinese civilian and military space programs. The visit gave impetus to the arguments of those who reject engagement.

The point is that communication between the United States and China on space issues has been limited at best. Consequently, there is a great deal of miscommunication, misinterpretation, misrepresentation, and poor assumptions made by each side about the other's intentions in space. In one regard, however, the message of the United States to China has been crystal clear: the United States is not interested in cooperative space programs with China. There have been several reasons for this attitude. Cooperation is associated with the term "engagement," a word summarily rejected by many conservative politicians as being associated with President Clinton's China policy. Some have linked space cooperation with tacit endorsement of China's stance on human rights or nonproliferation or want to link space cooperation to concessions from China in other areas, such as human rights or nonproliferation—linkages that did not work with the Soviets and likely would not work with the Chinese. Others refuse to work with a Communist country. Finally, the United States harbors concerns about transferring dual-use technology and worries that China's lack of reciprocity makes it a one-sided deal favoring China.

The United States says it is interested in working with China "as a global partner."[8] Yet actions have not matched words when in functional areas such as space, the United States maintains a strategy that it might characterize as hedging, but many see as containment,[9] as the United States tries to ignore the Chinese regarding space cooperation and isolate it regarding technology development. But trying to isolate China has proved impossible. It is convenient to assume that Chinese space technology has been acquired through begging, borrowing, and stealing, but more accurately, the Chinese have developed space capabilities by borrowing generic designs from others, participating in cooperative programs, developing technology indigenously, and buying what they needed and could afford from what others would sell them.

In a globalized economy, a plan to isolate China can be effective only if the United States has full control of whatever it wishes to deny, and there are few remaining areas in which the United States holds a monopoly. Space is one of the most globalized aspects of world commerce. Even if the U.S. military succeeds in dominating space, it is highly unlikely that the United States will ever be able to monopolize the space arena.

China has ambitions and generates concerns that go beyond those of many developing countries. Yet its development goals reflect those of other countries worldwide. Therefore, a closer look at Chinese space ambitions provides insight into both the types of activities and capabilities that all developing countries want now and in the future and the unique challenges and opportunities presented by the Chinese space program specifically.

Space and Development

Developing countries seek space capabilities for all the reasons discussed in chapter 7 regarding developed countries, such as building knowledge-based societies, technology development, and more. They desire satellite data networks for connectivity, for everything from pragmatically attracting more global information technology jobs to, for some, providing an infrastructure for the spread of democratic values. More immediately relevant to the governments in power in many developing countries, including India and China, large portions of the population are located in rural areas, sometimes extremely rural areas. Satellite communications link villages and cities, a priority for governments that recognize that historically, discontent has started not in the cities, but in the poorer, rural areas.

Developing countries also look to space for information that is essential in both the short and long terms. Bangladesh, one of the poorest countries on earth and prone to natural disasters, has upgraded its weather-forecasting capabilities through data from NASA satellites. Because so much of the country is lowlands, the flooding after a typhoon kills as many people as does the actual storm. With warning, people have the chance to leave the lowlands for safer terrain. South Asia also pioneered using satellite imagery to identify areas of ground soil densely saturated with water that could be breeding grounds for malaria-carrying mosquitoes, so that these areas could be sprayed with insecticide. Over the longer term, tracking changes in soil content, river beds, and alluvial plains can help governments plan for maximum agricultural use. Space-based information can also come in the form of telemedicine or tele-education. Doctors can read X-rays and CAT scans from a distance, allowing otherwise inaccessible expert advice to be included in diagnostic and treatment procedures. In tele-education, programs like the Pan-Pacific Education and Communication Experiments by Satellite (PEACESAT) use satellites to broadcast educational programs to remote island countries in the Pacific. The same premise that lured Europe into space over forty years ago—the idea that space leads to industrialization, and industrialization to development—has expanded in developing countries to reflect the importance of IT in the era of globalization.

The United Nations maintains the Office for Outer Space Affairs (UNOOSA), responsible for promoting international cooperation in the peaceful uses of outer space. The office serves as the secretariat for the General Assembly's Committee on the Peaceful Uses of Outer Space (UNCOPUOS). UNOOSA implements the United Nations Program on Space Applications (PSA) and works to improve the use of space science and technology for the economic and social development of all nations, but for developing countries in particular. Under the auspices of UNOOSA, the PSA conducts training courses, workshops, seminars, and other activities on applications and capacity-building in subjects such as remote sensing, communications, satellite meteorology, search and rescue, basic space science, and satellite navigation.

All in all, developing countries primarily view space assets as development tools. Most share the views of Europe, Japan, and many other countries that investing in technology with many potential benefits and uses is a boon rather than a detractor. While not necessarily looking to space assets to provide military benefits, the option—or even the perception of an option—is not unattractive.

China and Space

China does not pursue space activity for the sake of exploration. The pragmatic gains that space provides for economic development and attaining strategic goals are paramount:

> The Chinese government attaches great importance to the significant role of space activities in implementing the strategy of revitalizing the country with science and education and that of sustainable development, as well as in economic construction, national security, science and technology development and social progress. The development of space activities is encouraged and supported by the government as an integral part of the state's comprehensive development strategy.[10]

Manned spaceflight, usually associated with civilian activity and exploration, has become the public focus of China's activities. Like the United States, China has a longer military space history than a civilian one, but the civilian side is currently the more prominent. Having studied the U.S. playbook for the Apollo program, China recognizes that manned spaceflight, while costly, also yields multiple returns. The technology developed or improved through a manned space program can spill over into other areas, many associated with development.

Several parallels can be drawn between U.S. decision making in support of the Apollo program in the 1960s and China's in regard to its current manned space program.[11] Project 921, as the second Chinese attempt at a manned space program is called, was initiated and championed by former president Jiang Zemin.[12] It was undertaken in 1992 because the time was ripe. China was on an economic upswing and increasingly more technologically adept. It desired advanced space technology for both domestic telecommunications and the military, and it created a positive focal point for national pride to counter the negative images of the Tiananmen Square massacre.

On October 15, 2003, then-Lieutenant Colonel Yang Liwei lifted off into space from the Jiquan launch site, returning twenty-one hours later after sixteen orbits around the earth. *Shenzhou V* was followed two years later, almost to the day, with the October 12, 2005, launch of *Shenzhou VI*. That flight carried *taikonauts* Fei Junlong and Nie Haisheng on a five-day mission to low earth orbit. During their time in space, the *taikonauts* worked and lived in zero-gravity conditions, and established China as a country technically capable of extended space exploration and development.

But China did not send men into space because Jiang Zemin is a space visionary, eager to explore the heavens as part of human nature. Jiang is a pragmatist, a skilled politician, and a technocrat who ascended to power by maneuvering his way through the Byzantine maze of China's power structure. His support for the manned program was a calculated risk. Domestic pride, international prestige, dual-use technology development, and economic development, including skilled jobs and expanded science and engineering educational programs—all benefits that the United States reaped through Apollo—are proven reasons to pursue manned space programs. Jiang understood that space successes are spectacular, but so, too, are space failures. Not only were national goals on the line, but so was his own prestige relative to his successor as president, Hu Jintao. Both Jiang and Hu understand the credibility and legitimacy that they as individuals, the Communist regime domestically, and China internationally can reap from space.

The launch of *Sputnik* in 1957 was a huge psychological boost to both the Soviet people and the Soviet government during the Cold War and, conversely, a huge blow to both the people and the government of the United States. The pride generated in the Soviet Union from that psychological boost also translated into political credibility and, hence, governmental legitimacy for the Soviet Communist government. Credibility and legitimacy are important considerations to the government and its leaders in Beijing. One Chinese official said about the *Shenzhou V* launch, "This is not America where money comes from the taxpayers. This is money of the Communist Party—they would do with it what they decide. It is great they are investing into something that makes us proud."[13] Images of a Shenzhou vessel that makes people feel good about themselves and their country are found on Chinese consumer goods from phone cards to water heaters. A billboard in Beijing has a baby's hand holding a rocket; the slogan advocates family planning for a bright future. Beijing's interest in manned spaceflight for the domestic pride and international prestige it engenders parallels its interest in bringing the Olympics to Beijing in 2008. Yang carried an Olympic flag with him into orbit, ceremoniously revealing it on his return.[14]

Six centuries ago, Ming Dynasty inventor and Chinese national hero Wan Hu supposedly strapped rockets onto his chair and ordered his assistants to light them. When the smoke cleared, Wan Hu and the chair were, not surprisingly, gone. Yang Liwei, Fei Junlong, and Nie Haisheng have now joined Wan Hu as space heroes. Beyond NBA basketball star Yao Ming, China does not have many "living legends" or heroes, and so their importance in society should not be underestimated. Yang's bio at launch read like the proverbial right stuff: thirty-eight years old, a college-educated fighter pilot with a selfless wife and adoring son. He is described as having been a bright youth and a bit of a mischief-maker. In postlaunch interviews, he was personable, connecting well with average people. His political credentials must also be assumed to be impeccable, as he is both the new poster boy for the Chinese leadership and a role model for China's youth. A statue of Yang was immediately planned for his home province, Liaoning, a Rust Belt region ripe for the revitalization that Chinese *taikonauts* are intended to inspire. After the flight, the *Shenzhou V* capsule was displayed at the Millennium Monument in Beijing, where crowds estimated in the thousands had celebrated on launch day.

As opposed to the many celebrations that spontaneously erupted when Beijing was named as the city to host the 2008 Olympics, most launch-day celebrations in 2003 appeared to be largely choreographed. The celebration was directed and supported from the top levels of government, meant to be filmed and shown to the world as an advertisement of Chinese technical capabilities. Holding the largest celebration at the Millennium Monument, rather than in Tiananmen Square, also deflected comparisons with or reference to other times in Tiananmen that were neither celebratory nor reflective of national pride or unity.

In techno-nationalist terms, none of the other regional space contenders—including Japan and India—have managed the technical feat of launching people into orbit. The accomplishment carries with it a significant leadership connotation. Officials from around the world, and particularly Asia, sent congratulatory telegrams to President Hu in 2003. In India, space officials downplayed the technical aspects of China's launch: they confidently asserted that India could do the same if it chose to, but economics and need (what can a manned mission achieve that an unmanned mission cannot?) dictated that it should not.[15] Prime Minister Atal Bihari Vajpayee, however, congratulated China on its success and publicly encouraged Indian scientists to work toward a manned lunar mission. "Those who wonder what could be achieved by such space missions simply want the status quo to continue," he proclaimed prior to the launch.[16] It is unclear whom he was speaking to or about—the rest of the world, his own scientific community, or perhaps both. Just two days after China's *taikonaut* launch, however, India launched its most sophisticated remote-sensing satellite into orbit. The lack of subsequent fanfare certainly validated Beijing's manned spaceflight approach to garnering maximum prestige.

Initial Japanese responses to China's manned launch varied. Some space officials downplayed the technical significance of the event, while congratulating China. One Japanese official spoke directly in geostrategic terms: "Japan is likely to be the one to take the severest blow from the Chinese success. A country capable of launching any time will have a large influence in terms of diplomacy at the United Nations and military affairs. Moves to buy products from a country succeeding in manned spaceflight may occur."[17] One woman on the street was quoted in Japanese media coverage as saying, "It's unbelievable. Japan lost in this field."[18] While Japan's "losing" to China through Yang's launch was more perception than reality, China's success, juxtaposed with power failures on the *Midori-2*, a Japanese environmental satellite, and on *Nozomi*, a Mars probe, as well as the failure of two spy satellites to launch in November 2003,[19] resulted in calls to reexamine the Japanese program.[20] Space Activities Commission member Hiroki Matsuo candidly stated after Yang's launch that "discussions on manned spaceflight have long been simmering in Japan" and implied that Yang's flight would trigger Japan to reconsider its goals for space development. That reconsideration was initially thwarted by Japan's own launch difficulties. With the successful launch of its H-2A rocket in February 2005, media speculation immediately reemerged regarding a Sino-Japanese regional space race. Still, Japanese manned space activity remains mostly paper studies and speculation. While Japan has talked for years about a manned program, its consensus bound decision-making process, and a bureaucracy that moves at a glacial pace, creates almost insurmountable political inertia. Japanese officials admit that a budget increase on the order of two to three times the current figures would be necessary to undertake an autonomous manned program. Such an increase is not likely to occur.

Clearly, China has established at least the perception of being Asia's space technology leader, and perhaps the region's technology leader generally, even though Japan still creates more high-tech consumer goods. Both Japan and India are technically capable of manned programs if they muster and sustain the political will, but that is elusive in democracies. The high cost of manned spaceflight, and public prioritization of government spending on such programs as schools, roads, health care, and defense over space, makes space relatively expendable for politicians who have to make funding choices and be accountable to their constituents.

Internationally, by joining the United States and Russia in an exclusive club of countries capable of manned spaceflight, China has regained what it considers its rightful place among the world's technology leaders. But prestige alone, even in techno-nationalistic form, cannot sufficiently justify the expenditures inherent in a manned space program. It is difficult to estimate what the Chinese are spending on space exploration because of low labor and manufacturing costs compared with those in the United States and elsewhere. At a presentation at the Center for Strategic and International Studies in Washington in April 2006, Luo Ge, CNSA vice administrator, suggested that China's annual spending on space was about $500 million.[21] CNSA administrator Sun Laiyan later said that the Chinese government had spent approximately $2.4 billion on the first five Shenzhou spacecraft and that the budget for the first stage of the Chang'e lunar exploration program was just over $1 billion.[22] Best guesses are that China's actual space budget is about $2 billion annually. While that pales in comparison with NASA's annual budget of over $16 billion, and the more than $22 billion spent for U.S. military space programs, for China, it is a considerable sum. Pragmatic domestic returns consequently become as necessary as prestige.

Among their tasks as heroes, *taikonauts* are expected to encourage China's youth to pursue educational programs in science, engineering, and technical careers, giving them hope of someday being involved with the space program. In Japan, the "best and the brightest" university students are known to join companies based on recruiters' hype about the company's involvement in the space program. Though in reality they may spend their careers making washing machines, pride from association with companies involved in space efforts seems relevant in both education and career choices.

Education and on-the-job experience for the Apollo scientists and engineers created a generation of highly trained technical personnel. Engineering programs were specifically set up in U.S. colleges and universities in the 1960s to accommodate the need for new and specialized aerospace skills. Today, the

University of Science and Technology of China, Tsinghua University, Beijing University of Aeronautics and Astronautics, and Beijing Institute of Technology are among the top universities in China, and all eagerly discuss and promote their involvement in the space program. Student interest in space is said to have exploded in China after the *Shenzhou V* launch.

Education is important because a space program generally, and a manned program specifically, fits in with China's plans for economic development. The European approach of linking space to economic growth is also reflected in China's 2000 space white paper: "The Chinese government attaches great importance to the significant role of space activities in implementing the strategy of revitalizing the country with science and education and that of sustainable development, as well as in economic construction, national security, science and technology development and social progress."[23] The link between space and development is reiterated in the 2006 white paper, which says that China is aiming to maintain "comprehensive, coordinated and sustainable development, and bringing into play the function of space science and technology in promoting and sustaining the country's science and technology sector, as well as economic and social development."[24]

Education is a prerequisite to building an industrial base, and development in China requires jobs, preferably skilled jobs. When China began Project 921, it wanted to develop a cadre of trained engineers and scientists, and it has come a long way in that regard. China is proud that 80 percent of the workforce involved in that project is under forty years old, and many workers are under thirty.[25] China also realizes, however, that it stills ranks with other developing countries, such as Brazil and India, as having marginal capabilities in science. In that regard, it is still pulling itself up the learning curve, and it relies on space to play a large part in that effort.

Along with generating interest in science and technology as educational fields, space is, as it was for the United States during Apollo (and beyond), a jobs program. The China Aerospace Science and Technology Corporation (CASC), the organization primarily responsible for executing the manned program, employs over 150,000 people and has 130 subordinate organizations. The size of the Chinese space enterprise is not unusual for either Chinese strategic enterprises or space enterprises generally. At one time, an estimated 26,000 jobs in the United States, including NASA employees, contractors, and subcontractors, were estimated to be directly related to shuttle operations.[26]

China does not have congressional pork-barrel politics to contend with, as the United States does, but it does have a huge workforce to keep employed.

Many of the large Chinese state-owned enterprises (SOEs) are being privatized, and the SOEs are letting go of redundant workers who bloat the payroll. In response, a slow approach is being taken to balance economic efficiency with the need to keep people employed. The more experienced and skilled labor China can offer, the better the chances that the government can attract global industries and employ those who lost their government jobs to privatization. Boosting employment, attracting industry, and selling high-tech products and services, including within the aerospace field, are all Chinese priorities. A post-2003 launch comment from Yan Xuetong, a political scientist at Tsinghua University, reflects those priorities. "Now," he said, "people will realize that we don't only make clothes and shoes."[27]

Finally, China is most certainly developing space technologies for military use. Its 2004 white paper on national defense targets "information" as the orientation and strategic focus of military modernization. With space and information so closely linked, increasing its space assets and capabilities ranks high on China's list of priorities. China has seen the advantages that the United States military has reaped from space, and seeks to enhance its own position.

Development of the Chinese Program

Although the Soviets left an updated German V-2 rocket, renamed R-2, when they left in 1960, China developed its space program primarily alone, though not necessarily by choice. Mao Zedong scared off even Nikita Khrushchev with his casual attitude toward nuclear war. This was followed by self-imposed isolation during the Cultural Revolution, and then a tenuous-at-best relationship with the United States, especially after Tiananmen Square. Nevertheless, the Chinese have always maximized their ability to learn from others. Xinhua News Agency carried a comment in 2002 that "Russia's experience in space technology development was and is of momentous significance to China." It is also not a coincidence that the Xichang launch site is at approximately 28 degrees north latitude, while Kennedy Space Center is at 28.5 degrees north latitude. China picked a similar location to allow it to emulate U.S. postlaunch procedures and expectations, described in some detail and published in open-source U.S. literature. Even today, though the Shenzhou spacecraft bears similarities to the Russian design of Soyuz, technical comparisons seem to bear out that the Chinese version is China's own product:[28] rather than reinventing the wheel, China's designers improved the Soyuz design to make it their own work, viewed as simply a smart business practice. Chinese innovation is still significantly hampered, however, by its Soviet-style "bureaus and institutes" bureaucracy. China is trying to reform the system, but for the time being, it still labors under it.

The China Aerospace Science and Technology Corporation was created in 1999 to develop national defense and aerospace endeavors from the former China Aerospace Corporation (CAC).[29] To become more competitive, the Chinese government reformed the top defense and technology corporations, including CAC, a large SOE under direct supervision of the State Council. CAC and its roughly 270,000 employees were divided between CASC and the China Aerospace Science and Industry Corporation (CASIC).[30] CASC has a registered capital of 9 billion yuan (about $10 billion). Among the organizations subordinate to CASC are five large research academies—the Chinese Academy of Launch Technology (CALT), Chinese Academy of Space Technology (CAST), Shanghai Academy of Space Flight Technology (SAST), Chinese Academy of Space Electronic Technology (CASET), and Academy of Space Chemical Propulsion Technology—two large research and manufacturing bases,[31] and a number of factories, research institutes under the direct supervision of the headquarters, and companies in which CASC has major or minor shares. CASC employees include technical staff, researchers, and academicians from both the Chinese Academy of Sciences and the Chinese Academy of Engineering.[32] CASC management seeks to be globally competitive by becoming a modern enterprise system, with its own brands, intellectual property rights, and powerful core competitiveness.[33]

But reorganization and management reform to increase efficiency is complicated. In March 2006, China sentenced Li Jianzhong, former head of CALT and a top rocket designer, to life in prison for corruption and embezzlement of public funds. Zhang Lingying, the former CALT chief financial officer, was sentenced to twenty years in prison for helping him. To be able to compete in a globalized world, weeding out corruption in a system where corruption has been endemic will take time. China has to be creative, balance interests, and move slowly in its reform efforts. The aerospace industry has become somewhat of a test case for China.

The relationship between space and China's military is a complicated one. Like Russia and other countries, China did not initially bifurcate its program, as did the United States. It could neither afford to, nor did it want to. A single program was considered more efficient and better allowing for central control.[34] Although China has recognized the need to separate the military aspects of the program, directed by the Second Artillery Corps, from its civilian and commercial aspects, the specific steps taken so far have not satisfied either

Europe or the United States. Ultimately, however, the State Council, a civilian body, is responsible for setting policy regarding space. Under the State Council, CASC has general authority for manned spaceflight development and the Long March (Chang Zheng) series of rockets, and the CNSA is more or less NASA's counterpart. The Chinese military is responsible for functions such as security, logistics, and some (though increasingly fewer) facilities. All the *taikonauts* have so far come from the ranks of the military as well. The preeminent role of that State Council does not mean, however, that support from the military does not carry benefits to the program.

General Cao Gangchuan, while head of the General Armaments Department between 1998 and 2003, was the military commander for China's manned space program. In March 2003, seven months before the *Shenzhou V* launch, he was named one of the three vice chairs of the Communist Party's Central Military Commission, an even loftier position. That means he will very likely be even more able to advance his agenda, which has included support for manned space, though the State Council still ultimately can overrule him. General Li Jinai took his place as chief commander of China's space program. Cao's ascendancy illustrates an important point: part of the reason China has been able to follow through with a manned program, while other technologically advanced countries, such as Japan and India, have not, is the central control and support enjoyed at the very highest levels.

China did, however, follow the same model as the United States in launcher development. Marketed by the Great Wall Industry Corporation, the Long March launcher series bears a legacy not unlike that of the Delta, Atlas, and Titan commercial launchers, in that its original design in the early 1970s was for an ICBM (Dong Feng 4 and Dong Feng 5).[35] This ultimately unified effort in sensitive areas such as propulsion research and development gives rise to many U.S. concerns about technology transfer.

Among CASC's most important current achievements has been restoring the Long March success rate, or reliability, critical for commercial appeal. The lucrative Chinese launch market quickly dried up after the launch accidents in China during the 1990s and the subsequent issuance of the Cox Committee report in the United States. Other than launching twelve satellites for the Iridium communication venture, the commercial launch market, which had generated substantial hard currency for China, has been at almost a standstill, with little potential for a dramatic turnaround. The Chinese hope that positive spillover from a successful manned launch in terms of perceived technical capability—as of 2005, their reliability rate is over 90 percent—will benefit their commercial launch program. Nevertheless, an improved perception of Chinese technical capability will not change restrictive U.S. export laws or restrictive interpretations of ambiguous export laws, which prohibit the sale of U.S. satellites or satellites utilizing U.S. components to China.

The Chinese operate from three launch sites: Xichang, in Sichuan Province, for all geosynchronous satellites; Jiquan, in the Gobi Desert of Gansu Province, for Chang Zheng 2F (manned) and Shenzhou spacecraft; and Taiyuan, in Shanxi Province, for polar-orbiting spacecraft. The Iridium satellites were launched from Taiyuan. There has been speculation that China may build another site on Hainan Island, more open to the international community.

Besides launch vehicles, China has numerous satellite programs. Dong Fang Hong (DFH) communication satellites have gone through multiple iterations. *DFH-1*, also known as *Mao 1*, was launched in 1970 and is best known for broadcasting the song "The East Is Red" from orbit. China is developing the latest DFH iteration cooperatively with Germany. The Fanhui Shi Weixing (FSW) recoverable satellites were originally developed for photoreconnaissance, but now are also used for remote sensing. The third type of application satellite is the Feng Yun (FY) series, used for meteorology and remote sensing. China has also launched a series of Shi Jian satellites, carrying science payloads, and their regional navigation satellite system called Beidou. Some new satellites have been delayed. *SinoSat 2*, China's first direct-to-home (DTH) broadcast satellite, has been put off from 2005 to 2006 while relevant regulations for receiving and watching DTH programs are developed. But generally speaking, satellite program expansion is expected to proceed.

Manned Spaceflight

China's long march to manned spaceflight began in 1992. An earlier manned program, called Shuguang (Dawn), was started in the 1970s, but stopped in 1980 due to lack of funds, technological barriers, and a pragmatic decision to emphasize applications satellites. Statements made in 1996 planned for 1999 as the first manned launch, to commemorate the fiftieth anniversary of the founding of the Communist state. However, depressed finances—perhaps in part due to lost income from anticipated commercial satellite launches—and technical issues made it impossible to keep to the original timetable. There simply was not enough time for unmanned precursor missions to ensure that a manned flight would not meet with disaster. The flight of *Shenzhou I*, one of four unmanned precursor flights, occurred in November 1999, completing fourteen orbits and returning to earth after just twenty-one hours.

Like the Soyuz transport, the Shenzhou spacecraft has three primary modules and can carry three *taikonauts*. Both spacecraft have a service module, a command module, and an orbital module with a docking ring. The forward Shenzhou module is used to conduct experiments and will likely act as a crew transfer module in future Chinese space missions. These could include docking with another Shenzhou vehicle to form a rudimentary "space laboratory." Unlike Soyuz, the Shenzhou orbital module has a second set of solar panels, enabling it to remain in orbit independently for prolonged periods. Shenzhou also has a rear service module where the propulsion system is housed, with a manned command capsule between the service and orbital modules. In essence, China took a workhorse Russian design and upgraded it to accommodate China's needs and future plans. With no prior manned spaceflight experience, though, China bought selected Russian systems, including life support. In other cases, China built its own equipment, either because the equipment was too expensive to buy or because it wanted to better understand the fundamentals involved, or both.

Thus while Shenzhou bears an uncanny resemblance to Soyuz, differences are also apparent. Shenzhou is basically bigger and more technically sophisticated. The Chinese say it actually has more in common with second-generation spacecraft produced by both the Soviet Union and the United States, such as the Gemini spacecraft, than the first-generation Mercury or Vostok spacecrafts. This comparison is confirmed in the West.[36]

Between January 2001 and December 2002, China flew three more Shenzhou precursor missions. Each mission was progressively complex, testing life-support systems, spacecraft maneuvering, and landing. *Shenzhou III*, launched in March 2003, left the forward module in orbit. It carried a relatively sophisticated remote-sensing payload, a medium-resolution imaging spectroradiometer (MRIS) that transmits high-quality data to Chinese ground stations. The infrared technologies have potential civilian and military applications, again illustrating the inherently ambiguous nature of most space technologies, making it difficult to clearly determine the intent behind developing them.

Fourteen *taikonauts* were initially selected for training, and, much like the first U.S. astronauts, they were drawn from the elite ranks of military fighter pilots. The first two *taikonauts* were trained in Russia,[37] but the rest have been trained in a facility north of Beijing. Shrouding the training in secrecy, China deliberately created an aura of mystery and drama around its *taikonauts*, though after *Shenzhou IV*, it released footage of the *taikonauts* in training. The month before the *Shenzhou V* flight, three *taikonauts* were named as finalists for the honor of being the first Chinese into space. Yang Liwei was selected, though he was not informed until the night before the flight.

Again, the Chinese flew military equipment on the *Shenzhou V* capsule. Conservative analysts in the United States equated that with the program being a Trojan horse for military space activities. As Richard D. Fisher, vice president of the International Assessment and Strategy Center, testified: "And more ominously, the PLA may envision manned military space platforms inasmuch as its first manned spaceflight, the Shenzhou-5 of October 2003, was primarily used for military surveillance. It cannot be dismissed that future Chinese manned space stations planned for the next decade could perform defensive and offensive military-space missions."[38]

Either these analysts forget or do not care that the size of the space shuttle cargo bay was specifically dictated by the U.S. military so that the shuttle could carry intelligence payloads. Also missing from their analysis is that the Pentagon nearly stopped the International Space Station in its tracks by demanding the right to conduct research there. Consequently, these concerns about potential military use of manned Chinese space facilities come across as more than a little hypocritical.

Plans for the Future

Analysts tend to disagree only minimally on what China is doing in space, in terms of current lift capabilities, the types of satellites it is capable of building, and the number of launches to date. There is considerable difference in opinion, however, as to why China is engaging in space activities and to what end. Speculation thus runs rampant. Due to both the nature of dual-use technology and the size and complexity of China, evidence can be found for whatever thesis one seeks to prove. Analysts also disagree on how much of China's space programs are a reaction to the United States and how much are based on its own strategic ambitions.

Relatively speaking, China has been considerably more open about the Shenzhou launches, especially *Shenzhou V* and *Shenzhou VI*, than it has been about other space-related issues. Translating technological accomplishment into regional power requires such publicity about technological accomplishment; China realized that its manned launches had to be media events if it wanted to reap the rewards of its efforts and responded accordingly. But it is gaining confidence. *Shenzhou V* was broadcast with a time delay. *Shenzhou VI* was broadcast live, though Chinese journalists were warned in advance that they might have to hand over their videos if something went wrong. Foreign journalists have still not been allowed at the launches, though in June 2006, a closely chaperoned group of foreign journalists was taken on a tour of the mission-control center outside Beijing.

Part of China's propensity toward secrecy is cultural. When intent is ambiguous, failure is easier to avoid. Wanting to "save face" is an Asian trait, not just Chinese. Other Asian countries, however, have accepted the need for more transparency than China has, at least when dealing with an external audience. Further, Chinese secrecy laws, part of the Communist regime's attempts to control information, are so ambiguous that individuals involved in the space program feel that a circumspect approach—which translates to secrecy—is personally prudent. One never knows what seemingly innocuous information is covered by secrecy laws, and there are instances in which information has been declared secret after it has been innocently conveyed. Increasingly, however, there is internal public debate in China about its space ambitions, including questions about the money being spent on space when there are still so many poor in China.[39] In the United States, concern primarily regards the ultimate intent of Chinese space ambitions, but deciphering such intent can be difficult for outsiders. China's silence and secrecy have made speculation a primary source of information in the United States. There are other complicating issues as well, many regarding communication.

China is booming. Especially without dialogue, deciphering Chinese intent regarding space becomes considerably more difficult than surveying known capabilities. Analysis must be based on information from a variety of official and unofficial sources, with interpretations falling along a spectrum.

Open-source materials, particularly technical journals, are often used as sources of information about what the Chinese are working on or even thinking about.[40] Most technical journals are very technical, focusing on detailed discussions of optics, trajectories, and sensors. There is disagreement on how much can actually be gleaned from them. China analyst Larry Wortzel suggested that part of the difficulty with "intent analysis" is that "most technical articles from the science digests in China, admittedly, only deal in the theoretical aspects of how to fight war in space and analyze U.S. strengths and vulnerabilities."[41] Other analysts opine that "an aggressive pursuit of available open sources can yield a limited assessment of China's recent military modernization ambitions and achievements"[42] or that there is a wealth of information in the technical literature from which inferences about possible intent can be drawn, including about ASATs, if carefully translated and followed over periods of time.

Recently, there has been considerable concern in the United States about low-tech and largely legal Chinese efforts to gather know-how and bits of information about space programs. These efforts include the people-intensive process of painstakingly sifting through mountains of open-source American technical literature, as well as Chinese visitors collecting information through casual discussions, conferences, workplace knowledge, and theft.[43] The value of information that can be gleaned from open sources should not be underestimated. According to the Defense Science Board (DSB), open-source material is among the most useful and least expensive collection options for China.[44] Perhaps the problem is not only that the Chinese do too much open-source analysis, but that the United States does not do enough or do it well enough.

Beyond technical journals, the volume of information and analysis that is produced in China and commercially available is increasing exponentially. A wider range of opinions tolerable to the government are appearing within academia and in the media. There are now both official publications, which are vetted by the government, and commercial, unvetted publications. Media outlets are proliferating, driven by market competition. However, whereas Americans understand the risks of relying on the *National Enquirer* or a lone blogger for "facts," the need for similar discrimination among Chinese sources does not always seem to be understood by U.S. analysts.

The statement by Chinese analyst Wang Huncheng used as an epigraph for this chapter is perhaps one of the most often cited Chinese quotes on "intent," even though Wang is not necessarily an authoritative source. Similarly, while Americans understand that a treatise on defense policy from a university professor or a War College student being encouraged to "think outside the box" does not necessarily reflect U.S. government policy, they do not have the same understanding of Chinese writers in similar positions. A Chinese source widely heralded by American conservatives as indicative of policy is *Unrestricted Warfare*, written in 1999 by two colonels at a Chinese military institution. While it reveals a line of thought among at least some officers, it does not necessarily reveal official policy or even mainstream thought regarding Chinese intent.[45] The increasing information available from China from numerous sources increases the potential for communication misfires. That being the case, careful source checking by analysts is imperative.

The Department of Defense's *Annual Report on the Military Power of the People's Republic of China* for both 2003 and 2004 contained references to Chinese "parasite" satellites for potential use as ASATs. The 2004 report stated that the claim was still being investigated. According to Union of Concerned Scientist researchers Gregory Kulacki and David Wright, a relatively easy

Internet search in China places the origin of the story about those satellites with a self-proclaimed "military enthusiast" named Hong Chaofei from a small town in Anhui. Since its posting on the Internet in October 2000, the story has appeared and been cited multiple times. Hong's Web site also contains scores of stories on "secret" Chinese weapons to defeat America in a war over Taiwan. China is working on small satellites, but the parasite satellite appears to be more one man's fiction than fact.

There are other instances of misinterpretation as well. *Challenges to Space Superiority*, published by the National Air and Space Intelligence Center at Wright-Patterson Air Force Base in March 2005, highlighted quotes from Zhan Liying of the Langfang Army Missile Academy suggesting that China will "threaten on-orbit assets." Kulacki and Wright again tracked down the quotes and the source, and again found several key errors in both translation and interpretation of context, fully documented in a published Union of Concerned Scientists research paper on Chinese military space capabilities.[46] Key words were omitted from the actual Chinese quote, and there were misinterpretations of what was included. For example, "should," indicating a recommendation about a decision not yet made, was misinterpreted as "will," indicating what China intended to do or was doing. Further, the author was found to be a junior faculty member at a facility primarily responsible for live-fire and simulated training for junior artillery officers, where ASAT research was likely not even taking place and that subsequently has been shut down. Again, not exactly an authoritative source for U.S. government planning purposes.

That said, China is working on a wide variety of dual-use research potentially applicable to ASAT development, including microsatellites and small satellites. Some of this research is cited in DOD's *Annual Report to Congress on the Military Power of the People's Republic of China* for 2005, though not always accurately. The medium-resolution earth observation Tsinghua series being built with Surrey Satellite Technology Limited of Great Britain is included, though the resolution for *Tsinghua-1* is stated as 40 meters when it is actually 30 meters—information easily found on the Internet. Meanwhile, the *Naxing-1* is not mentioned in the report, though it is in many ways more interesting as a totally Chinese effort with some sophisticated upgrades. It is currently the smallest satellite, with three-axis stabilization. Its purpose is stated as "high tech experiments." Chinese commitment to commercial small satellite development, for applications including mapping and environmental monitoring, is further evidenced by the December 2004 opening of a microsatellite industrial park in Beijing, a commercial venture with over 170,000 square feet of floor space. That venture is not mentioned in the DOD report either.

Also, Chinese space supporters, politicians, and even officials state what they would like to see happen in interesting and perhaps provocative ways that do not reflect official U.S. policy—just as American politicians do. Unfortunately, the U.S. media sometimes do not interpret them that way. The media have reported that China would put a man on the moon by 2010, establish lunar bases, be mining helium-3 on the moon, and have a base on Mars by 2040. CNN once reported the cancellation of the Chinese manned lunar program, even though an official manned lunar program has never been announced.

The Chinese are not immune from mistakes either. China's *Global Times*, known for its strong anti-American bias, reported in December 2004 about a supposed 1990 launch of a U.S. "hovering" satellite that could get within 30 centimeters of satellites to jam them or, if necessary, smash them. In both cases, rumor was reported as fact. In other cases, Chinese scientists and space supporters are deliberately floating their own ideas to the Chinese press to test the domestic political waters, trying to determine what interests Chinese politicians. Their statements are misread as fact. And while the Chinese manned space community has enjoyed the support of the government, it still finds itself facing the same types of questions that NASA gets from politicians in the United States; it must justify its existence in the face of other, more pressing needs for government expenditures. Like their counterparts in the United States, Chinese space scientists find themselves focusing not on what should be done toward a rational manned exploration program, but what will sell politically.

China, however, can be its own worst enemy in exacerbating opportunities for misunderstanding and nefarious assumptions about its intents. The problem is part technological, part cultural, and part political. Communicating with Chinese space officials can be difficult at best. Office phone and fax numbers often do not work or are not answered. Voice mail is rare, and e-mails regularly bounce back to senders. Cell-phone numbers are worth their weight in gold, as people actually answer them. Information regularly released in the West is often held closely in China, because information generally is more closely held due to cultural norms. Until recently, rarely have Chinese officials talked about their own program at international conferences, and if they do, presenters usually stick to officially approved and thinly transparent scripts. There have been a few welcome exceptions to this rule. At the first conference of the International Association for the Advancement of Space Safety (IAASS), held in Nice, France, in October 2005, a CASC official openly invited international cooperation on Chinese programs. A CNSA delegation headed by Luo Ge, in presentations in Washington and at the annual meeting of the National Space Society in Colorado in April 2006, was also refreshing in its openness. If China wants to be a member of the international family of spacefaring nations, such transparency, open discussion, and encouragement of differing perspectives should become the norm.

Communication with China can be problematic, especially for those outside the business community. Business calls, however, are promptly returned, and faxes are answered when they are received, as there is money at stake. European officials also say that they have few problems beyond occasional technical issues. For Americans outside the business community, however, there is a definite chill that is political in nature. China would likely respond that its communication problems are just an unexplainable phenomenon—akin to Chinese citizens' difficulties in getting visas to visit the United States, as their applications are regularly and arbitrarily rejected, which they complain about often and loudly.

Officially, China's plans for its manned programs are phased, incremental, cautious, and ambitious. China will proceed as long as politicians see enough return on investment to justify expenditures, whether payback is in terms of regional techno-nationalist clout, domestic government legitimacy, technology, jobs, education, or other factors. Money being a constraint, China will focus on high-payoff areas in which it can match or pass others. Manned space programs have appeal for their techno-nationalistic benefits: satellites and robotics for economic development. Satellites are particularly important, as they provide domestic linkage with rural areas, particularly in western China; communication capabilities important to attract foreign investors; capabilities to monitor and manage environmental problems and disasters; and force enhancement capabilities for the military. China's plans to increase its military space capabilities are also ambitious, though not even close to the level of U.S. ambitions.

China has officially announced a three-step plan for its manned space program: *taikonauts* in space, a space laboratory, and, eventually, a space station. The first phase of the first step was completed with the success of the *Shenzhou VI* mission. The second phase, likely to commence in 2008, will include more advanced technical maneuvers, including docking and space walks. The Shenzhou design allows the orbital module to operate independently, so the module and a *taikonaut* could be left in orbit as a rudimentary space laboratory. More ambitiously, China will likely send a small laboratory, of perhaps eight tons, into space. Because the orbital modules have docking rings, they could also be hooked together as a space station. But China clearly intends something more than two docked modules as its space station. A mock-up of a twenty-ton version, 4 to 5 meters in diameter, has already been built and exhibited. That manned space lab could be launched around 2015. In the meantime, China has incremental and doable options.

China expects its subsequent manned flights to be successful, but it is acutely aware that failure is possible. Especially after the *Columbia* accident, discussions in China focused on that possibility and with the need to nevertheless move forward. Perseverance is seen as part of the heroic struggle required of spacefaring nations—though undoubtedly that perseverance would not be without political limits.

Regarding lunar exploration, in 2003, China announced its unmanned Chang'e program (named after a mythical Chinese fairy), consisting of lunar orbiters, lunar rovers, and a soil-return spacecraft. The first spacecraft in that program is scheduled to be launched by 2007, followed by a soft landing in 2012, and lunar sample return in 2015.[47] At the meeting of the International Lunar Exploration Working Group held in Beijing in July 2006, Chang'e was among one of several international robotic lunar exploration programs included in what was proclaimed the "lunar decade" of exploration. The program is intended as a precursor to a manned lunar program if political interest, technical success, and funding can be sustained. Ouyang Ziyuan, head of China's lunar project, who often makes statements used as unofficial trial policy balloons to the domestic and foreign media, stated as much in November 2004. It was likely not coincidental either that shortly after NASA announced that its "Apollo on steroids" program would put a man on the moon by 2018, Ouyang goaded that "China will make a manned moon landing at the proper time, around 2017."[48] Chinese space officials are adamant, however, that if they continue, their program will build infrastructure for sustained use, rather than planting a flag or returning a moon rock—a reference to the United States abandoning its manned lunar program and failing to use it as a step farther into space. Whether that adamant position can be sustained if challenged by politicians, as has often been the case regarding who dictates NASA's direction, will be interesting to watch. Based on how long it took China to make its first launch—about eleven years from approval to first flight—it would likely take it somewhere between six and twelve years to achieve a manned mission to the moon, once approved. Others believe that it can be done much faster.

If China develops an extended manned program, it might consider several options. While there is no evidence of this occurring, one option suggests that if China were to engage in a covert manned lunar space race with the United States, and won, it would be a techno-nationalist public relations coup, worth the effort and money. A manned lunar landing would require an operational Long March 5 launch vehicle, which is still in development. But a Chinese manned flight that orbits the moon potentially could be completed using current technology, though China would need experience in docking and maneuvering techniques that it has still not tested.[49] If such an event were to occur, say, in conjunction with some other event—like the 2008 Olympics—it is not at all clear whether global citizens would differentiate between a lunar orbit and a lunar landing before assuming that China had done what the United States was unable to do. Such a scenario would require total secrecy from China to prevent a U.S. challenge and a considerable amount of chutzpah. The argument might be made, however, that such a scenario fits with a Chinese strategic approach based on the concept of *shi*. Prevalent in Sun Tzu's *Art of War*, *shi* uses deception and intelligence to create opportunities for surprise.[50] Again, there is no evidence that such a scenario is currently being pursued, but it is worth watching for evidence to the contrary.

Apart from a lunar destination, which would be a repeat of Apollo, the Chinese might try for something different, but only marginally more difficult, such as a manned mission to a near-earth asteroid—two or three potential candidates are identified every year—to obtain samples. These samples would be of significant interest to both the scientific community and the commercial sector (for ore and mineral content) and would mean an achievement beyond any by the United States. As a "first," the perception of one-upmanship over the United States would be perpetuated and render China even more bang for its buck. Interestingly, if China decided on such a mission within the next three to five years, and if the same time line could be assumed for a near-earth asteroid mission as for a lunar mission, China might reach an asteroid at about the same time that the United States was returning to the moon. If current—and, many feel, unrealistic—U.S. schedules are not met, China would have the manned heavens to itself.

To go to the moon or a near-earth destination, or even to launch a space station of any substantial design and consequently weight, the Chinese will need a new launch vehicle. Future launch vehicle designs provide for increasing lift capabilities, using a concept similar to the U.S. Evolved Expendable Launch Vehicle. A family of vehicles is being created based on one design, with a range of capabilities. The lift capabilities planned and being developed are of such a magnitude that they are obviously intended to support such missions as Chang'e, a space station, a manned lunar program, and potentially a Mars program as well. Successfully developing the new Long March 5 launch vehicle, officially expected to debut in 2008, is a prerequisite for many of China's more ambitious plans. In November 2004, Chinese media reported that a new heavy-lift engine had been successfully ground-tested, but experts say it will likely be closer to 2010 before the new launcher is available. The new engine would increase Chinese lift capabilities from two to fourteen tons to GEO and twenty-five tons to LEO, which is what China will need for the Chang'e mission.[51] Even more lift will be needed for a space station and potential manned lunar programs. These powerful launchers also expand Chinese capabilities to launch heavier military satellites.

Military Space Plans

China's goals for expanding its military space capabilities are less clear than they are for its manned space program. From outside China, Chinese silence is equated with hiding something or nefarious intent; the Chinese aversion to transparency has perpetuated American policy being based on worst-case speculation.

Just as is the case with European countries, Japan, India, Pakistan, Russia, and others, China intends to develop space-based force enhancement capabilities as part of its military modernization efforts. In recent years, Chinese warfighting doctrine has shifted in many of the same ways as has American warfighting doctrine, away from comprehensive war with large platforms to smaller wars under modern, high-tech conditions. Wars of the future are envisioned in Chinese doctrine to be more limited—that is, not nuclear—shorter; more destructive; still decisive; fought on land, in the air, at sea, in the electromagnetic spectrum, and in space (as related to information technology); and emphasizing service cooperation, or jointness. Furthermore, China expects to fight a technologically sophisticated opponent. Largely from observing U.S. successes, China understands that to conduct such a war effectively, it must be a space power. China is under no illusions that it can control space all the time, nor does it feel that it has to. It needs only to buy the time it needs to accomplish its goals by interfering with its opponent's capabilities. China's goals focus first and foremost on keeping Taiwan from declaring independence.

Chinese doctrine states that thwarting the capabilities of opponents can be achieved through "soft kill" activity[52]—that is, interfering with information systems and ground stations by various means, including electromagnetic pulses,[53] or passive counterspace, such as camouflage, flares, and deception. Enhanced communication capabilities for command and control, and imagery and gathering information from space, are all on the agenda of the People's Liberation Army (PLA). Seeing space as zero-sum, the United States has not been pleased by Europe's plans to build autonomous space systems for military use, let alone by China's expansive space plans. And just as it is both unrealistic and impossible to stop European countries, Russia, or Japan from proceeding as their budgets and political wills allow, it is also unfeasible to stop China.

The escalation of China's military space program is based on more than modernization, though. Beijing's view of the heavens is not the same as Washington's; indeed, when Beijing looks into space, its view is largely obscured by U.S. assets. The actions, assets, and rhetoric of the United States give China even more acute discomfort than Europe, Russia, Japan, or others feel.

Two events in 2001 were critical from the Chinese perspective. First, the United States issued the report of the Space Commission, which stated that space would inevitably become a battleground, so the United States would be remiss not to prepare—the unspoken assumption being that preparation meant developing space weapons.[54] Second, the United States held its first-ever space war game, called Schriever I.[55] In that well-publicized game, U.S. forces were pitted against an opponent threatening a small island neighbor about the size and location of Taiwan. It did not take China long to conclude that it, in turn, would be remiss not to prepare for the inevitability of the United States developing space weapons, especially as China might be the target. Since then, the United States has held two more space war games. Schriever III, held in February 2005, saw a 350-person team of space professionals engaged in battle in a global environment scenario set in the year 2020. Who, the Chinese ask, was the United States battling against?

Politically, China has also observed the rise of the Blue Team as a major influence in U.S. politics. While Sino-American relations have warmed in some areas since September 11, Taiwan remains a potential flashpoint in the near term, and the potential for China as a peer competitor to the United States is a consistent concern of those who view zero-sum competition as inevitable. Subsequently, both actions and rhetoric have convinced China that the United States is developing both hardware and doctrine to use against China, potentially even preemptively.

Chinese military and political literature on space and warfare over the past ten years shows a clear pattern of concern about the actions and words of the United States about its space ambitions. Numerous Chinese reviews of specific U.S. space weapons initiatives look at technology being developed without a specific mission stated, such as the "Rods from God." This concerns China greatly, as it understands that U.S. acquisition law supposes that technology is developed for particular missions. Chinese authorities then begin to speculate about what those missions might be. As Chinese literature also emphasizes Chinese inability to compete with American space technology for the foreseeable future, the question for China then becomes what it must do to retain its sovereignty and freedom of action on issues of critical national interest, such as Taiwan, if China faces the advantages the United States has because of its space assets.

China understands that space provides considerable force enhancement capabilities to the United States, force enhancement that could deter China in the case of conflict over Taiwan. China is not necessarily looking for a high-tech, high-end response to the advantages of U.S. space assets (as the United States would), as both funding and capabilities are limited. This returns to the premise that all threats to the U.S. are asymmetric. Taking out a satellite ground station is easier than taking out a satellite, and with more deniability and less risk from an anticipated "proportional response."

Whether China has an active antisatellite program remains ambiguous. It has an active microsatellite program, and the U.S. Air Force is developing microsatellites for potential ASAT purposes. China probably does not want to spend the money on a robust ASAT program, but it will if it feels that it has to, such as if the United States were to deploy space weapons considered threatening to Chinese sovereignty. As chapter 5 discusses, China has repeatedly called for a treaty with Russia banning weapons from space, unquestionably with the self-serving purposes of saving itself the money of having to develop an asymmetric response. If China developed an ASAT program, it would be technologically limited. Chinese tracking abilities currently do not have global reach, though China's capabilities in this area have been improved through its manned program. Nor does China have a "launch on demand" capability requisite for a truly robust ASAT program. Finally, China's space infrastructure is relatively vulnerable due to its stationary nature; it would be easy for the United States to dismantle it.

Manned to Military?

What has China's military gained from its manned space efforts? One set of benefits is relatively indirect. In an October 21, 2003, article in *People's Daily*, Zhang Qingwei, deputy commander of China's manned space project and president of CASC, gave specific information about China's rocket and capsule.[56] He said that China had achieved breakthroughs in thirteen key technologies, including reentry lift control of manned spacecraft, emergency rescue, soft landing, malfunction diagnosis, module separation, and heat prevention. Earlier Chinese publications had cited additional areas of technical advancement, including computers, space materials, manufacturing technology, electronic equipment, systems integration, and testing. Spacecraft navigation, propulsion, and life support were specifically cited for potential application to dual-use civilian–military projects.[57] The Chinese upgrade of the Jiquan launch site and its entire tracking system will certainly benefit the Chinese military, as will any additional upgrades to its tracking capabilities.[58] Moreover, the Chinese military will benefit from experience in areas such as in-orbit maneuvering, mission management, launch on demand, miniaturization, and computational analysis. Experience extends not just to building hardware, but to program management and integration as well.

While both the United States and the Soviet Union initially tried but were unable to identify any military advantages to a man in space rather than unmanned systems,[59] China seems determined to explore that premise for itself, likely through the use of the orbital module at some later date. *Shenzhou III* left its orbital module aloft in March 2002,[60] as did *Shenzhou V* and *Shenzhou VI*. At some point, China may leave a *taikonaut* in orbit for a period of time. Clearly, China is intent on getting the maximum return from its investment and will explore all potential uses of the Shenzhou hardware.

Cooperation versus Competition

On a regional level, in 1992, China, Pakistan, and Thailand formed the Asia-Pacific Multilateral Cooperation in Space Technology and Applications (AP-MCSTA) group. Intended to actively promote regional cooperation and establish a regional space cooperation initiative, the group transformed into the Asia-Pacific Multilateral Cooperation in Space (APSCO) organization in January 2003. More than fourteen countries attended the second meeting of the drafting group for the APSCO convention. In the meantime, a Memorandum of Cooperation on Small Multi-Mission Satellites (SMMS) and Other Projects was signed in 1998 and 1999 among Bangladesh, China, Iran, Mongolia, Pakistan, South Korea, and Thailand. With the United Nations Economic and Social Council for Asia and the Pacific, the group has also sponsored two short-term training courses on space technology and remote-sensing applications. All this organizational activity looks impressive, but APSCO's actual activity level has been relatively low.

As mentioned before, China has also worked with Brazil on the high-resolution electro-optical imaging satellites. It has also worked with the European Space Agency on the Double Star space science program, to measure the effects of the sun on the earth's environment, and the Dragon program, in which European and Chinese researchers are working to monitor and forecast Chinese air quality. Of most pressing interest to the United States, however, has been Chinese involvement in the European Galileo program, because it has been feared that the already demonstrated high value of American GPS navigation satellites to the U.S. military would accrue to China through Galileo. The Chinese Beidou regional navigation satellite system is based on 1980s technology. Its actual capabilities, such as whether it potentially includes weapons targeting or is primarily suitable for civil systems such as transportation, has been subject to interpretation. Clearly, however, access to Galileo would offer Beijing a significant capabilities upgrade. Of more recent concern is apparent Chinese commitment to building a navigation satellite system, Compass, not as a regional system, but along the lines of GPS or Galileo.

China has signed cooperative space agreements with a number of countries, including Canada, Germany, Italy, France, Britain, Russia, Pakistan, India, and Brazil. The scope of cooperation includes developing the *Dong Fang Hong 3* communications satellite with Germany, a broad Russia–China agreement, and narrow scientific co-ventures. One especially interesting future area of potential international cooperation is launch services, since participation in international launch services was included as a Chinese white paper goal.

On the commercial side, in December 2004, China landed the contract to build Nigeria's first satellite, *NIGCOMSAT-1*. China will also launch it in 2007. China's bid, against competitors from the United States and Europe, was competitively priced, offered a launch date only two years from contract signing, and included job training for fifty Nigerians. With this contract, China joins the United States and Europe in being able to offer a full package of satellite manufacturing, launch, and servicing. It also positions China to take the lead in working with other developing countries, where job training in technical fields always sweetens a deal. China used the same commercial model in 2005 to secure a communications satellite contract with Venezuela. The satellite, to be named *Simon Bolivar* after the South American revolutionary, will be launched in July 2008. As part of the deal, ninety Venezuelan specialists will be involved in the satellite's development.

Finally, the Chinese have built a cooperative arrangement with the University of Surrey Space Centre in Great Britain, much to the consternation of the United States. Having built and launched over twenty-five microsats that perform a wide range of scientific missions, including earth surveillance, Surrey has specialized in marketing this new capability to developing nations. Its customers include Chile, Malaysia, Taiwan, Egypt, Algeria, Nigeria, and China. As chapter 5 discusses, microsatellite technology is a concern because of its potential value as an ASAT. China has alluded that it might consider using this technology to deny the United States the use of space in a crisis or conflict. How it would do that is not known, nor is it known whether China is deliberately working on an active ASAT program or is merely developing all the requisite pieces should it decide to take that route.

In December 2004, China announced the creation of a national engineering and research center for small satellites to develop large-scale production capability. While widely reported that China wants to build six to eight small observation satellites each year and have over a hundred in orbit by 2020, other interpretations of Chinese press reports suggest that China would contribute only eight satellites to the more than a hundred in the Geostationary Operational Environmental Satellite network by 2010. The stated purpose of these satellites is to create a large surveying network of Chinese territory, for monitoring water reserves, forests, and farmland. But it is all dual-use technology.

Though China has coveted inclusion for some time, it is not a partner on the ISS, a program known as much for its political aspects as for its technical utility or capability. Inclusion is symbolic of acceptance into the international family of spacefaring nations—a sign of legitimacy for the government in Beijing—and could bring China up the technology learning curve. The ISS partners have been expected to contribute technology, money, or both, and until recently, China has not had either. In 1997, however, the international consortium of partner countries, led by NASA, offered Brazil, a country with far less space experience than China, the opportunity to provide flight equipment and payloads, as well as a flight opportunity for Brazilian astronaut Marcos Pontes to the ISS. This makes it more difficult to dismiss the premise that China's exclusion is at least partly political. And while it can be argued that because it is an international consortium, it is not up to NASA to offer China participation in the program, China is aware that a unilateral offer from NASA without consultation with the partners is exactly how Russia joined the program in 1993.

Conservative politicians in the United States have objected to including the largest remaining Communist country in the world in a program largely designed to show that countries could peacefully work together. Media comments from Representative Dana Rohrabacher (R-Calif.) in 2001 regarding discussions about increasing international financial contributions to the space station are illustrative. While acknowledging that China might have the resources to contribute to the station, Rohrabacher said that he ruled out approaching Beijing due to that country's human rights abuses: "The space station's supposed to stand for something better."[61] The question that must be asked, however, is whether the benefits of exclusion outweighs the costs already cited.

At a July 27, 2005, hearing before the House Armed Services Committee, two views on the benefits and dangers of contact with the Chinese, particularly the military, were expressed and are fairly typical of those prevailing. Franklin Kramer, former assistant secretary of defense for international security affairs, spoke in favor of contact:

> I think that if we use the right public information we can make sure that we have the Chinese understanding really what we're about. We can also try to get a better understanding of what they're about. They're non-transparent, I think, would be a kind word. And we have sometimes tried to get really reciprocal visits. We have not achieved reciprocal visits. But I think we can nonetheless get some good insights by going there and talking to their people and getting as much as we can.[62]

As Kramer states, engaging with China allows information to be gathered from its still largely closed society. Further, creating Chinese dependence on American technology offers the United States more leverage than does pushing Beijing closer to others. Including China in cooperative manned programs also utilizes Chinese funds that might otherwise go into military programs, making the Cold War space-race scenario plaguing the United States vanish, and emphasizing U.S. leadership in a positive manner.

Richard D. Fisher expressed a different view:

> When China does launch a space station, I think we have to consider that that space station may serve both military as well as civilian purposes. And when we look at our own potential future cooperation, dialogue, space dialogue with China, we have to keep this in mind. That when we invite—if we were to invite—a Chinese astronaut onto the space shuttle, that the information technology that that single individual might pick up could be turned into a potential Chinese military space platform.[63]

The suggestion that serious technology transfer might occur if a Chinese *taikonaut* were to participate in a shuttle mission is astonishing. The United States tried to share technology with other developed countries in the formative years of Communications Satellite Corporation (COMSAT), even making blueprints and manuals available; it was found to be very difficult. Further, while there appears to be concern that China will develop a significant manned military capability, history shows that both the Americans and the Soviets tried to find an advantage to a manned military presence and could not. The Manned Orbiting Laboratory, a program planned by the Air Force to house military astronauts, was canceled. Sensors have much better eyesight than do astronauts. Is there a fear that Chinese ingenuity will be able to find value in a military man in space that eluded the U.S. military? As a 2002 RAND report stated, "No compelling reasons for military human presence in space have been identified and few interesting ones have been conjectured."[64] Therefore, there seems little basis for fears regarding Chinese military men in space gaining some strategic advantage on the United States.

China's desire to be involved with the ISS has been evidenced both technically and politically. Technically, Shenzhou uses a Russian ASAS-89 docking mechanism, which is also currently used on both the shuttle and the ISS to allow for docking with both American and Russian spacecraft. Chinese officials continue to raise the issue of potential ISS participation at their infrequent program presentations and through second- and third-tier channels. If Chinese interest in the ISS has recently waned, it is more likely due to now-tenuous U.S. commitment. With limited funds, China may not be interested in putting money into a project with a limited future. With Russia and the other partners still pushing full station use, China is likely hedging its bets.

The United States has historically viewed international space cooperation as both a political carrot and a technical way to shape other countries' space activities. While the notion of shaping can be considered an intrusion on a country's sovereignty, if approached toward providing a win-win outcome for all parties, it can be both doable and desirable. Taking a different path with China has likely, though inadvertently, contributed to China's determination now to become a space power.

A United States–China Space Race?

China has no aspirations to match the military capabilities of the United States generally, or in space specifically. It has no illusions in that regard. The U.S. military budget is eight times larger than the Chinese military budget, making parity a highly unrealistic goal. From China's perspective, that leaves asymmetric strategies as the rational and perhaps the only choice. Clearly, using space assets against the United States or targeting U.S. space assets is not an option that any country takes lightly. The possible ramifications are substantial and deadly.

The view from Washington of Beijings's space activities focuses on three points. First, China's increased ability to use dual-use assets as force enhancers, as part of its military modernization, is troubling. Second, its acquisition of technology with potential use as weapons, specifically in the area of ASATs, is worrisome. Third, the perception that China is "beating" the United States in a space race is annoying, though most officials deny even thinking about it.

As stated before, the United States is unequivocally ahead of all other countries in space technology. There is no contest on the military side of the equation, hence the references to potential asymmetrical responses. Whether economics in another decade will continue to support that premise is not clear. In manned space programs, however, China's current political will to support an incremental program, even if it launches once every two years, can create the perception that it is "beating" the United States and perpetuates a Cold War view of space.

When the United States gets motivated, there is no stopping or beating it. But when it loses its momentum, there is the opportunity to be passed by others who simply take a slow, incremental, but consistent approach. With two manned launches, China achieved a significant engineering feat—but that significance should not be overstated. Launching a rocket is difficult; it requires attention to literally thousands of minute details. Launching manned rockets is even harder, as evidenced by the fact that only three countries have done it. But manned rocket launches do not equate to a major scientific breakthrough, such as those of the World War II–era Manhattan Project, in which scientific breakthroughs were an integral part of success. China has not leapfrogged over the United States by any stretch of the imagination, though people seem to be willing to believe otherwise.

Continuing along the current policy path allows a space race between China and the United States to occur through several scenarios. First, it has been suggested, usually by members of Congress pursuing these arguments to generate money for NASA, that deliberately engaging China in a space race would provide the political support for the U.S. manned program to advance, just as the Soviet Union's activities did for Apollo. They also argue that a space race would trigger a spending spree in China, with effects similar to those experienced by the Soviet Union trying to keep up with the SDI program, to the detriment of China and the benefit of the United States. Both are flawed analogies. During the Cold War, two competing superpowers started from the same point technologically and engaged in an engineering race. Both were motivated to compete. China has no reason to "race" the United States; its spending will not increase to keep up with or outpace that of the United States either, as China fully understands that it is impossible. Perpetuating a race with the Chinese will not motivate Congress to support the kind of funding needed to get the United States back on a leadership path either. There is simply not enough money to support all the military space programs and manned space programs, too—especially while supplemental budget requests of over $100 billion are being requested for war efforts in Iraq and Afghanistan. Space programs are so expensive that decisions to fund them usually boil down to classic guns-or-butter decisions. If the goal is to get more money allocated to NASA and civilian space programs, perpetuating Cold War–style race scenarios will likely backfire, and result in more going to military space programs.

In a race driven by perceptions, there is another angle to be considered as well. Chinese space capabilities, manned and unmanned, offer options to other countries not previously available. China could combine its capabilities and funding with those of other countries' capabilities to counter U.S. space dominance and avoid dependence. Working with China also offers other countries the opportunity to be partners, perhaps even take the lead on space projects, rather than being a participant, as the United States offers. Further, working with China offers other countries an option at a time when the commitment of the United States to its partners is not perceived to be particularly high. A sustained manned program supported by China and other countries in could leave the floundering U.S. program looking like an also-ran.

A Cold War version of a space race today would be a space arms race, but one in which the United States would actually be racing against itself to keep ahead both offensively and defensively. In such a race, for each new technology the United States creates, China would only have to find a way to counter it. To do otherwise would risk its own security interests, such as Taiwan. The counter need not be high-tech. Then the United States will have to come up with an even larger and more complex technological solution. This is the security dilemma:

> The first world is the worst for status quo states (offense has the advantage/offense-defense posture is not distinguishable). There is no way to get security without menacing other, and security through defense is terribly difficult to obtain. Because offensive and defensive postures are the same, status-quo states acquire the same kind of arms that are sought by aggressors. And because the offense has the advantage over the defense, attacking is one of the best routes to protecting what you have; status quo states will therefore behave as aggressors. The situation will be unstable. Arms races are likely. Incentives to strike first will turn crises into wars.[65]

The developmental benefits offered from space mean that it will be impossible to deny space technology to developing countries that can afford it. The United States does not have a monopoly on the technology that other countries seek. The imperative for connectivity to avoid being roadkill in the globalized economy makes access to the benefits of space a matter of national security that countries will not be denied. The inescapable fact that 95 percent of all space technology is dual-use creates problems.

China's determination to regain what it considers to be its deserved place in regional and, by default, global politics will provide significant impetus for continuing its space activity. Wanting to minimize the technology gap with the United States, to the extent that China can provide an asymmetric challenge in matters of its sovereign interests, is considered imperative. Generally speaking, China feels that it would be remiss not to prepare for what it sees as inevitable: the development of space weapons by the United States. Further, all Chinese foreign policy is about Taiwan, over which China is likely willing to fall on its own sword if threatened. Taiwan is the one issue that strikes at the very legitimacy of the Beijing regime, which feels that a successful Taiwanese bid for independence might threaten not only the future of one-party rule in the People's Republic, but the stability of mainland China itself as a unified state. There are likely to be at least some reckless elements in the Chinese leadership—particularly within the senior ranks of the Chinese military—who would be willing to risk nuclear confrontation with the United States rather than accept the collapse of the party's authority and the fragmentation of the Chinese state, an outcome that to some Chinese leaders might be just as bad as a nuclear exchange. Unfortunately, both China and Taiwan feel that their sovereignty is at stake, but China in particular is determined not to become the twenty-first century's version of the Soviet Union, exploding into flinders as each secessionist or political opposition group, emboldened by the Taiwanese example, increasingly rejects the authority of the central government.

The benefits of space-related activity, civilian or military, makes interaction between the United States and China on space issues inevitable. The only question is whether it will be preponderantly competitive or whether Washington will consider cooperating with China as well. The best option is always to use a full range of approaches to act in the best interests of the United States. If the United States remains recalcitrant about not working with China in space, China will find other countries to work with and continue to develop its program nevertheless, and the security dilemma will continue to escalate. This can be avoided. The question is whether there is the political will in the United States to actively change the current course of action.

NINE

Avoiding a Clash of Ambitions:

Toward a Comprehensive U.S. Space Strategy

And the stars of heaven shall fall, and the
powers that are in heaven shall be shaken.

—Mark 13:25

Much of the American public views space as an interesting museum exhibit, and Congress largely ignores it unless constituents' jobs in their district are at stake. Yet space has become an integral part of everyday life, not only for those in the United States, but for individuals all over the world. Space-based commercial navigation capabilities have evolved into a global utility; though they are currently available only through the GPS, other providers are likely to appear in the future. People will not be denied—or even take the risk of being denied—the services that these navigation satellites provide. The GPS is an American program and another example of the United States leading the way into space as it did with the Apollo program. The effect of Apollo on everyday lives was perhaps less explicit, but just as powerful.

When U.S. astronauts walked on the moon, an interplanetary "us" and "them" was inherently created, "us" being the people of earth and, if you believe that space is too big a place to have only one populated planet, "them" potentially still to be found. No other person will ever have the same role in history as Neil Armstrong. He was the first person from earth to step onto another celestial body. He, an American, led the way for all humanity. He demonstrated to people everywhere that no dream was too big and thereby dared them to dream as well.

Global leadership has characterized the role of the United States in space, and it is part of the conception of America as the "shining city on the hill." Forgoing that leadership dims the lights in the city at a time when the place of the United States in the world must be more pronounced and positive than ever before. Fighting the (long) global war on terrorism and shaping the world into a more stable place where the needs of human security are provided for all requires both U.S. leadership and global cooperation.

The United States needs a comprehensive space strategy that simultaneously addresses its leadership in manned space exploration, maintains a healthy aerospace industry to accomplish its goals, and protects its security interests. There are times when strategy is best executed through a number of separate policy initiatives, but this is not one of them. The current mismatch of strategies and resources, and the failure to consider the unintended consequences of particular policies and programs, are the results of that approach, mistakenly assuming that it is possible to separate parts of issues that inherently overlap. The new National Space Policy,[1] a sweeping document intended to provide guidance for America's multiple space programs, continues the country on its current, disjointed path, with shifts from the 1996 policy coming mostly in tone and priorities. The new tone asserts the unhindered rights of the United States in space, rather than acknowledging the rights of all nations.

The policy was released late in the afternoon of October 6, 2006, a Friday before a three-day weekend, by the White House Office of Science and Technology Policy. The document had been signed by President Bush on August 31, but deliberately released more than a month later with as little fanfare as possible—thus continuing the established government approach of maintaining a low profile to avoid too much scrutiny and, consequently, potential controversy.

At first blush, the National Space Policy of the Bush administration appears much like that of the Clinton administration,[2] which it supersedes. Upon closer examination, however, and reading it in the context of actions taken over the past six years, the changes are dramatic—as well as ambiguous and sometimes inconsistent with existing policies and programs. Most important, the policy enunciates an American space program that is first and foremost about military security. While not necessarily inappropriate, it does support the idea that the United States views space activity through a lens different from that of most other spacefaring nations.

The background section of the policy states that "those who effectively utilize space will enjoy added prosperity and security and will hold a substantial advantage over those who do not." Hence the importance of space is recognized—indeed, have- and have-not-country status based on effective space utilization is implied—but the realization that therefore all nations will seek and expect full utilization privileges seems not to be in evidence in much of the rest of the document. In fact, the next sentence states, "Freedom of action in space is as important to the United States as air power and sea power." That assertion not only focuses on the security (that is, military) aspect of utilization, but raises a question that can be raised elsewhere in the document as well: Are similar rights and expectations considered legitimate for countries other than the United States? While one might assume that if the United States claims a right or expresses an expectation for itself, it implicitly grants the same right or recognizes the same expectation for others, the language of the National Space Policy does not explicitly say that and, in fact, at times infers otherwise.

For example, one of the principles stated to guide American space programs and activities is that "the United States rejects any claims to sovereignty by any nation over outer space or celestial bodies, or any portion thereof, and rejects any limitations on the fundamental rights of the United States to operate in and acquire data from space." If no nation can claim sovereignty over space, why doesn't every nation have a fundamental right to operate in and acquire data from space? Another principle states that "the United States considers space systems to have the rights of passage through and operations in space without interference. Consistent with this principle, the United States will view the purposeful interference with its space systems as an infringement on its rights." Is it implied that other countries have the same rights?

The very first principle—and one cannot overlook the possibility that priorities can be inferred from order—says that "the United States is committed to the exploration and use of outer space by all nations for peaceful purposes, and for the benefit of all humanity. Consistent with this principle, 'peaceful purposes' allow U.S. defense and intelligence-related activities in pursuit of national interests." National interests are, of course, whatever the powers-that-be deem them to be. Before September 11, 2001, the national interests of the United States were configured hierarchically: from defense of the homeland as critical to survival, through economic well-being as vital and favorable world order as serious, to promotion of values as desirable. After September 11, all these interests have been on an equally critical footing. Linking "peaceful purposes" to national interests more closely connects space to defense and the military than ever before.

The Air Force's request in May 2005 for explicit rather than implicit approval in the National Space Policy to move ahead with the development and deployment of space weapons resulted in more attention being paid to the subject than the White House wanted. Most of the scrutiny remained in policy circles, but the potential for it to spill over into the public arena proved enough to preclude the overt approval of weaponization. Consequently, ambiguous language is employed where explicit language might have drawn the ire of Congress or the public. For example, in the discussion of the rights regarding the space capabilities the United States considers vital to its national interests, the policy states that the United States preserves its rights, capabilities, and freedom of action to "dissuade or deter others from either impeding those rights or developing capabilities intended to do so." Given the dual-use nature of space technology, what capabilities might be specifically targeted? The document goes on to say that the United States has the right to "take those actions necessary to protect its space capabilities, respond to interference, and deny, if necessary, adversaries the use of space capabilities hostile to U.S. national interests." That language is not new, but in its current context, it takes on more of a right of preemption than of defense.

The policy explicitly reiterates the four military space missions and support for missile defense, and then restates that the secretary of defense shall "develop capabilities, plans, and options to ensure freedom of action in space, and if directed, deny such freedom of action to adversaries." While the phrase "space weapons" is never uttered, it can be heard if one listens closely.

Other new, noteworthy areas of emphasis appear in the policy as well. It specifically rejects, for example, new legal regimes or other restrictions that would inhibit American access to space. It recognizes the need to train space professionals, addressing the kind of workforce issues raised by George Abbey and Neal Lane.[3] It also states much needed support for—hopefully, improved and expanded—space situational awareness: the ability to know what is happening in space, especially near one's satellites. Somewhat curiously, the policy gives considerable space to support for space nuclear power. This likely refers not to the large propulsion systems of the past, which were used to propel spacecraft to the outer solar system and beyond, but to specialized equipment used in smallsats and microsats, which have considerable potential military value. The need to improve development and procurement issues that have resulted in space projects routinely being overbudget and behind schedule is addressed, as is that for more interagency space partnerships. The latter can be expected to yield more NASA–DOD partnerships, which is likely the only way NASA's budget will be increased. On that note, the policy offers support for a "human and robotic program of space exploration" but fails to note that NASA's budget is not adequately funded to accomplish the goals set by the administration for travel to the moon, Mars, and beyond.

Gratefully, international cooperation is given a nod, with the secretary of state taking the lead in "diplomatic and public diplomacy efforts, as appropriate, to build an understanding of and support of U.S. national space policies and programs and to encourage the use of U.S. space capabilities and systems by friends and allies." Unfortunately, the more those friends and allies hear about the new policy, the less inclined they might be to work with the United States.

On October 11, 2006, less than a week after the new National Space Policy was released, Robert Luaces, alternate representative of the United States to the United Nations General Assembly First Committee, made a statement about the NSP:

> The international community needs to recognize, as the United States does, that the protection of space access is a key objective. . . . It is critical to preserve freedom of action in space, and the United States is committed to ensuring that our freedom of space remains unhindered. All countries should share this interest in unfettered access to, and use of, space, and in dissuading or deterring others from impeding either access to, or use of, space for peaceful purposes, or the development of capabilities intended to serve that purpose.[4]

What Luaces apparently did not realize, however, is that many countries regard the United States as the biggest obstacle to unhindered access to space.

Finally, Luaces presented a summary characterization of the space landscape from the U.S. perspective: "One: there is no arms race in space. Two: there is no prospect of an arms race in space. Three: the United States will continue to protect its access to, and use of, space." The same perspective was reiterated by an administration official, who spoke on the condition of anonymity: "This policy is not about developing or deploying weapons in space. Period." He further stated that new arms control agreements were not needed because there is no space arms race.[5] Skepticism remains.

The National Space Policy of 2006, even more than those before it, is disconnected from considerations of the ways that other countries might react to or the unintended consequences of the policy itself. Because the United States does not have a monopoly on developing space capabilities, a comprehensive U.S. space strategy must be developed considering the goals and capabilities of other countries. Not doing so, as Alfred Kaufman pointed out, puts a perpetual burden on the United States to keep outdoing itself technologically, because others can counter U.S. achievements with less technical skill and far less money. As Robert Jervis noted, it also keeps escalating the security dilemma (as discussed in chapter 1), a situation in which parties are drawn into conflict even though neither desires it. Finally, policy created in a vacuum also misses an opportunity to build U.S. soft power with other countries, which is critical to meeting our own needs and enhancing the stature of the United States in the world.

Developing a comprehensive space strategy for the United States, one that would still stress space security but on a broader basis, would take rigorous analysis by people from many disciplines and areas of interest. This has not been done yet, and in some areas seems to have been deliberately avoided. It would also require the National Security Council; the Departments of Defense, State, and Commerce; NASA; other government organizations; aerospace industries; academia; advocacy groups; and others to work together. It would take real government leadership.

The rationale for Congress to stay the funding course would have to come from the executive branch, not just with an announcement, but with a willingness to prioritize the program. Initially, funding for the Apollo program was "fenced," meaning that it was not subject to the kind of bureaucratic and congressional horse-trading common in the budget process. The executive branch would also have to sell the program, reminding the U.S. public of its importance. Based on the past, public buy-in cannot be assumed or overlooked. Space security through leadership would be the intended goal of a comprehensive space strategy, and it must be effectively conveyed as such to the public.

The book *Space Security 2004* defines space security as "secure and sustainable access to and use of space, and freedom from space-based threats."[6] That publication provides assessments of trends and developments in space based on survey results from global space experts. It also acknowledges that "U.S. experts have a tendency to view space security developments differently from EU, Chinese, or Russian expert participants."[7] All security is a matter of perspective. So far, how U.S. security is best served in and from space has been a question left to too few people, and from a very narrow military perspective that created the security dilemma in the first place. Space security must be redefined in the United States to ease the tension of the security dilemma and preserve American space leadership in the military, civilian, and commercial domains. The recommendations offered toward achieving those multiple goals are not all-inclusive. Rather, the intent is to suggest the kinds of issues that need to be considered.

Ameliorating the Security Dilemma and Maintaining Military Space Leadership

Thomas M. Nichols argues that Ronald Reagan's initial strategy against the Soviet Union was one of "overdoing it."[8] In both actions and rhetoric, Reagan made his offensive against the so-called Evil Empire clear. That included the Strategic Defense Initiative, dubbed "Star Wars," the legacies of which are now in silos in Fort Greely, Alaska, and Vandenberg Air Force Base, California. The rationale during the Reagan years was that if SDI caused the Soviets to pause for one minute before firing a nuclear weapon, then it was worth whatever it cost, as a nuclear exchange between the United States and the Soviet Union would certainly not have stopped at one nuclear weapon being fired; annihilation of both populations was the more likely outcome. While SDI eventually died under the weight of its own cost, it was an initially effective as part of a larger strategy to scare the Soviets.

After a while, though, the Soviets worried about ideological extremists making their way into the White House, and worried further that such extremists and what the Soviets regarded as their somewhat insane ideas were becoming the majority. The Soviets came precariously close to crossing a line from a useful and healthy fear of the United States to a counterproductive assumption that U.S. policy makers were unhinged, supporting arguments for Soviet preemptive nuclear strikes against the United States. There was also domestic concern in the United States about Reagan's campaign, with many Americans afraid of the direction their own government was taking.

Luckily, Reagan realized the counterproductive nature of going too far, and in the late 1980s, the United States made an abrupt turn away from competition and confrontation, toward cooperation. Reagan will certainly not go down in history as dovish, unpatriotic, or wimpy; he knew, however, when to change direction to best serve U.S. national interests. The United States today may be precariously close to crossing a line again, away from engendering healthy respect and fear and toward creating the perception that American policy makers are simply irrational. That perception will serve hardliners and radicals in other countries well. And while some might like to dismiss outside perceptions of U.S. policies, in fighting the global war on terrorism in Iraq, the United States has certainly found that it ignores such perceptions at its peril. Many countries feel that the United States does not consider others in foreign policy making and implementation.[9] When it comes to U.S. space policy as a subset of foreign policy, based on weaponization attitudes and the politics of the International Space Station, that view has been repeatedly validated.

The Russians and the Chinese have been seen as both susceptible to the irrational rantings of fear-mongers who say that space weapons are coming and perpetuators of propaganda on the same topic. Nevertheless, what some dismiss as merely "bold rhetoric" by the Air Force with little substance behind it,[10] others view differently. As chapter 8 discusses, analysts in the United States take

Chinese "bold rhetoric" seriously. Why would the United States expect anyone else to do otherwise? It is time to step back from the bold rhetoric, if that is all it is, and focus on how to best protect the national interests of the United States.

It is also time to be clear about the intentions of the United States. The *Report of the Defense Science Board Task Force on Strategic Communication*, published in 2004, suggests that "strategic communication describes a variety of instruments used by governments for generations to understand global attitudes and cultures, engage in a dialogue of ideas between people and institutions, advise policy makers, diplomats, and military leaders on the public opinion implications of policy choices, and influence attitudes and behavior through communications strategies."[11] The United States is lacking in its strategic communication about space in a number of areas, particularly regarding China.[12]

Rather than the "Space Pearl Harbor" analogy that appears to broadly drive American attitudes toward Chinese space activities, perhaps another analogy should be considered instead: "Space 1914." While far from a perfect analogy, the image of two countries becoming locked in a particular understanding of a strategic environment and unnecessarily setting themselves on a course for future crisis with considerable escalatory potential does fit. The resultant conflict could perhaps be wholly avoidable, if the participants had a better understanding of the true situation. Better strategic communication is required to prevent history from repeating itself, and engaging in a dialogue of ideas between people and institutions, according to the definition of the Defense Science Board Task Force, is one of the four fundamental premises of strategic communication. But the United States has to date summarily rejected that premise regarding China and space. The message from the United States has been clear in that regard. Whether it is the right message, however, is increasingly doubtful.

If the United States seeks to use strategic communication to influence decision makers and general populations—with influence defined as the ability to shape or affect others' beliefs and actions—then engagement appears necessary, though not necessarily sufficient. Engagement is also necessary to learn about China. Misunderstandings are better avoided through direct communication than inferences and speculation based on sometimes less than credible sources. A number of potential avenues can be pursued to begin dialogue, which could lead to greater transparency and expanded cooperation.

When General Richard Myers visited China in 2004, he announced the intent of the United States to pursue greater bilateral military relations, which up to that time had been fairly nominal. Donald Rumsfeld's visit to Beijing in October 2005 advanced that policy. General James Cartwright, commander of the U.S. Strategic Command (Stratcom), has expressed interest in establishing closer ties with both China and Russia on space issues, specifically to reduce the potential for misunderstandings. In fact, he has suggested that an exchange of field-grade officers might be one approach to forming better relationships.[13] Greater military–military contact could reap considerable rewards for the United States. Perhaps most important, some of the opacity of Chinese programs could be penetrated, allowing for more accurate analytic considerations of Chinese intentions in space. Also, better military–military relations could lead to an environment more conducive to backing away from the zero-sum attitude toward space that creates the security dilemma.

One of the results of NASA administrator Michael Griffin's trip to China in September 2006 is an agreement for officials from CNSA and NASA to meet annually to discuss potential areas of cooperation. There is considerable work to be done in merely beginning to understand how to effectively work with China. At the highest level, the United States has attempted to convey to Chinese elites a general willingness to work with them. Translating that general willingness to work together into meaningful dialogue in functional areas has been problematic because at the functional level, the United States tends to want to delve into specifics uncomfortable for Beijing. In United States–China defense consultation talks and military maritime security, for example, the United States seeks transparency on specific capabilities, deployment, and spending, which China avoids. For its part, China is more interested in engaging in functional area dialogue to better understand U.S. strategic intent on issues such as support for Taiwan, the military alliance between the United States and Japan, the nuclear ambitions of North Korea, and space. Consequently, even when dialogue infrequently occurs, both sides can end up frustrated by lack of progress on their goals. Part of strategic communication is learning how to create an effective dialogue.

Traditionally, space cooperation has often started small, in nonthreatening areas such as space science, and grown when confidence is established between partners. Environmental monitoring and astronaut rescue also offer mutually beneficial, relatively nonthreatening areas of potential cooperation. Scientists are always willing to work together on goals of mutual interest and benefit to all mankind, providing a good place for cooperation to start, out of the way of politics. This was evidenced in the United States quietly providing limited information about space debris to the Chinese before the launch of *Shenzhou VI*. Nobody wants to see astronauts or *taikonauts* endangered.

While not the norm, there is also precedent for official U.S. cooperation with China. Chinese scientist Guo Huadong was the only Chinese representative on a team of forty-three scientists working on the Shuttle Radar Topographic Mission (SRTM), a program to map the world in three dimensions, flown by the shuttle *Endeavor* in 2000. Guo, director of the Remote Sensing Application Research Institute under the Chinese Academy of Sciences and an expert on radar remote sensing, also worked in the *Endeavor* radar scientific working group between 1991 and 1996. Ironically, the SRTM mission was not pure science; the product of the mission was very practical and dual use, capable of being employed or purposes ranging from environmental and crop monitoring to targeting. The mission was cosponsored by NASA, the National Geospatial-Intelligence Agency (NGA) of the Department of Defense, and the German and Italian space agencies.

The Chinese also actively participate in international groups, such as the Committee on Space Research (COSPAR) and the International Lunar Exploration Working Group (ILEWG). Chinese space scientists engage in exchanges with American universities and, even more frequently, with universities outside the United States. Cooperation does occur; for the most part, it has simply been outside the official, government realm.

Besides incremental moves toward cooperation, there is also the option of taking a more dramatic step, such as flying a Chinese *taikonaut* on the shuttle and perhaps even to the ISS for a quick visit. There are benefits that could be rendered by such a gesture.

With an authoritarian government in place, Chinese public opinion does not have the power it has in the United States, but it is increasingly becoming a force with which the Chinese leadership must contend. A full spectrum of attitudes toward the United States can be found, as evidenced in a study by the Pew Global Attitudes Project, conducted in June 2005.[14] Clearly, however, the Chinese are influenced by single events. Chinese citizens reacted virulently to both the accidental U.S. bombing of the Chinese embassy in 1999 and the death of the Chinese pilot in the EP-3 incident over Hainan Island in 2001. If the Chinese are negatively affected by events, perhaps they can be positively affected, too.

The current U.S. approach to strategic communication seems to understate the importance of positive "singular opportunities" and images, though the increase in favorable opinion toward the United States after its tsunami relief efforts in 2004 clearly demonstrated that opportunities exist. A single bold act, such as allowing a Chinese *taikonaut* on a shuttle flight, could generate a powerful, positive effect on Chinese public opinion. Such a shuttle flight would produce tangible images and news coverage much the same as the

Apollo-Soyuz program did in 1975. If a goal with strategic communication is, in part, to alter Chinese public opinion, these images could be very potent.

Globally, the message that the United States is trying to send regarding its intentions in space is unclear. The United States seems almost to have multiple personalities when it denies any intentions to develop space weapons, on the one hand, and lets Air Force officials boast of their accomplishments and flashy programs in that very area, on the other. Further, holding and widely publicizing a space war game with China as the obvious "enemy" could be interpreted as indicating U.S. plans. Was that the intent?

The United States must decide what message it wants to send to China and other countries about space, and do so clearly and consistently. That effort would be very useful in alleviating the security dilemma. If the United States intends to use space hardware for purely defensive purposes, the dilemma created by the difficulty of distinguishing between the offensive and the defensive nature of space technology could be relieved by a no-first-use policy from the United States. The overwhelming dominance of U.S. assets means that any attempt to interfere with those assets would be damaging but not crippling, and would certainly ensure the full wrath of American power in return if such interference was attempted.

China is interested in developing military space capabilities as part of its military modernization effort, as are most countries in the world. It is further interested in developing space capabilities as part of globalization efforts and to send a techno-nationalist message regionally and globally. But China is also responding to the message it hears from the United States. We must make sure what we intend to convey is what is being heard.

Moving Toward Strategic Stability

Strategic stability occurs when countries maintain a defensive advantage, making offensive movements imprudent. Currently, the United States has and is seeking to maintain an offensive technological advantage, regardless of the implicit or explicit path to reach that goal. Relying exclusively on technology for security—in this case, space weapons—does not provide an asymmetric advantage; it creates a strategically unstable environment. Even in 1978, Jervis pointed out that "the security dilemma is at its most vicious when commitment, strategy, or technology dictates that the only route to security lies through expansion."[15] Currently, technology is driving strategy in an ultimately counterproductive manner.

Developing space weapons is also, as pointed out throughout this book, horrifically expensive. Continuing on the current path of development not only exacerbates the security dilemma, but puts manned space leadership in jeopardy, because classic guns-or-butter decisions become inevitable. Cold War images of a space "race" succeed only in generating more spending on military space programs to counter the "threat," rather than generating momentum to re-create the Apollo program.

Further, through clumsy rather than intentionally nefarious use of its considerable power, the United States is perceived as a rogue nation in it own right. Other nations regard the United States as skirting international law in its treatment of war prisoners, lack of support for international treaties, and proclivity toward preemption and unilateralism. In the space arena, movement toward space weapons further reinforces this perception. The commitment of the United States to a regime in space based on legal premises and parameters would demonstrate its commitment to the rule of law at a time when that commitment is doubted, and when it is dearly needed to support U.S. efforts to spread democracy and principles of good governance. A new paradigm must be considered.

The United States should adopt a military space strategy that refocuses its efforts back to space support and force enhancement in the near term, and away from active forms of space control, including missile defense and force application. Quite simply, the benefits that accrue from such a shift outweigh the risks. With limited funding for space activities generally and military space activities specifically, providing maximum support to the warfighter through space support and force enhancement offers the United States the most bang for its buck and, ultimately, greater space security. As part of this approach, a threat-based risk analysis is inherently needed, based on facts rather than seeking confirmation of a predetermined, ideological conclusion. Such an analysis is imperative to decide the levels of effort in research, development, deployment, and funding that are appropriate among the multitude of programs available to be pursued. Refocusing U.S. efforts to space support and force enhancement is not enough, however. Additionally, the United States must solidify its position of strength by codifying the status quo.

Formal, legally binding agreements such as treaties can be difficult to negotiate and ratify, but they are sometimes the best way to manage complex security issues. In the case of space, they appear to be the only way to break away from the security dilemma currently defining the space environment. However, rather than focusing on the same kinds of frustrating arms control efforts of the past, which tried to control ambiguous dual-use technology and created difficulties that justified forgoing further arms control efforts, these efforts would focus on establishing parameters for acceptable space activity. In conjunction with technical, economic, diplomatic, and military efforts, such measures could provide more security for the United States in the long run than could measures using technology only. Using all the policy tools available to the United States rather than relying primarily on technology "fixes" allows the United States to avoid a costly arms race with itself, codify the current status quo with it as the clearly dominant power, and step away from the hair-trigger security dilemma that not only prevails, but is escalating. On a broader level, it also allows the United States to lead by example, which it has always done better than any other country, rather than by force. Many kinds of legal arrangements can be considered.

Confidence-Building Measures

Confidence-building measures (CBMs) are a time-honored way to deal with security apprehensions. They are most often used with strategy or policy changes in which one or both parties perceive considerable risk. When changes were made in post–Cold War nuclear strategy, "Washington took the initiative by announcing a shift in nuclear doctrine, negotiating strategic force reductions, and introducing confidence-building measures that were intended to reduce tension and foster relations."[16] The same must be done in space.

Surveillance is an important confidence-building measure that has not received sufficient attention and, consequently, funding. Gratefully, both the new National Space Policy and Strategic Command commander General Cartwright support increased attention to surveillance. More and better information is needed about what is going on in space. Space surveillance provides inspection capabilities, and, according to Jervis, "by relieving the immediate worries and providing warning of coming dangers, inspection can meet a significant part of the felt need to protect oneself against future threats, and so make current cooperation more feasible."[17] If anything has been learned in the post–September 11 world, it is the value of solid information. Therefore, better situational awareness in space through upgrading and expanding the Space Surveillance Network must be a U.S. priority. It is also critical that the United States preserve its commitment to transparency, as it requires a similar commitment from others.

To be sure, the issues involved with space surveillance are complicated. One that has been repeatedly raised in this book has been the difficulty in determining intentions, American, Chinese, or otherwise. Space surveillance offers concrete information about what is happening in space. Yet not only are space surveillance assets limited, but what is available is becoming increasingly restricted.

Since the 1960s, NASA has published data through the SSN's Space Control Center, making it free and available to the public through its Web site. The data is widely used by amateur sky-watchers, academic researchers, industry groups, and foreign governments, but recently, it has been increasingly restricted. In 2003, legislation was passed stating that the secretary of defense must preapprove all users of the data, and those approved are not allowed to redistribute it.[18] The latter provision could prove difficult to enforce and lead to more restrictions. The net result, however, is that there is increasingly less transparency regarding activity in space—and just as a lack of transparency in China heightens U.S. concerns, so too does restricting U.S. transparency raise concerns outside the United States, especially when accompanied by bold rhetoric. Consideration must be given to the benefits of access to information, as well as to the risks, and a workable solution must be found. When the United States becomes more opaque regarding space issues while insisting that the Chinese open up, it smacks of hypocrisy.

Rules of the Road

As has been done on land, at sea, and in the air, defining what is acceptable and what is not in space is both prudent and necessary. What constitutes a threat? How close does a spacecraft have to move toward another spacecraft before it is considered a hostile act? Defining threats and establishing rules allows actors to plan for potential responses to those who break the rules. Nefarious actors often do not play by the rules, but rules provide parameters for identifying nefarious actors as such. Taking action against the space asset of another country would draw worldwide public attention. Therefore, it is essential that the United States be able to define and validate claims of hostility against its space assets according to recognized international parameters.

Rules provide parameters of action, and various groups have been working on developing rules of the road for space for some time. The Center for International and Security Studies at the University of Maryland, for example, in collaboration with the Committee on International Security Studies at the American Academy of Arts and Sciences, is engaged in the Reconsidering the Rules for Space Project. The project's goal is to explore "operational rules and transparency measures that could be used to balance the interests of commercial, civilian, and military constituencies in the United States and other major space-faring countries."[19] Another group at the Henry L. Stimson Center has developed *A Model Code of Conduct for the Prevention of Incidents and Dangerous Military Practices in Outer Space*,[20] which provides examples of areas where defining parameters of operation might be useful, such as

- "Special caution zones" around spacecraft: areas around spacecraft off-limits to others without prior notification and approval
- Agreement that parties not simulate space attacks by any methods
- A noninterference rule for satellites
- Development of safer "space traffic" management procedures, based on International Telecommunications Union rules

There is already precedent for such strictures, such as the 1972 Soviet-American Agreement on the Prevention of Incidents on and over the High Seas. Since the military compares space with the air, land, and sea in other ways, creating rules of the road for space should not be too far a stretch. In discussing possible areas of cooperation on space policy with Russia and China, General Cartwright expressed interest in formulating joint standards for satellite station keeping, including routine maneuvering, to ensure that the military of one country is not alarmed by the sudden maneuvering of the satellite of another country and does not assume that it is the beginning of an attack. This is exactly the kind of activity envisioned by those interested in the development of rules of the road.[21]

Develop the Principles of the Outer Space Treaty

Categories of technologies could be established, with some designated as permissible and some prohibited. The Outer Space Treaty prohibits orbiting weapons of mass destruction. Expanding on the concept of "peaceful uses" of space in the OST, a protocol could include other categories of weapons as well. Part of the challenge would be to define what was to be prohibited, given the difficulty in defining what constitutes a space weapon due to the dual-use nature of much of the hardware involved. A first step might be prohibiting certain "active" space-based systems, such as hypervelocity rod bundles or shooting systems, such as lasers.

Taking a legal approach to ensuring space security does not mean forgoing other options. The United States can adopt a hedging strategy toward space weapons development, not crossing any provocative lines, but retaining the technical ability to build and deploy hardware if circumstances change and the need arises. In the meantime, however, the United States has more to lose in space than anyone else, due to both the number of its space assets and the consequent reliance on them. The United States has never been in a stronger position militarily in space relative to the rest of the world, but that position cannot be maintained through technology alone.

Unfortunately, between fears about U.S. intentions to weaponize space, constraints on American companies' abilities to act as reliable and rational aerospace business partners, and the United States potentially backing out of international commitments like the ISS, the U.S. leadership image has taken a beating. America is a leader. It is strategically in our interest to remind other countries of the positive aspects of our strong leadership capabilities, and manned space offers an opportunity to do so.

Manned Spaceflight: A Leadership Opportunity

Does the leadership image of the United States need a makeover? It is sometimes more important to be feared than loved, but it appears that the United States is precariously close to overdoing it, if it has not already. Tanks, planes, and lasers will not stop the spread of feelings or ideology. And whether the United States likes it or not, a poor image clouds any positive, progressive message that we want to project. For the United States to lead in the long term, it must have willing followers.

In the 1960s, leadership was the motivation that took the United States to the moon, as the country wanted to show itself as the winner in a technology-based competition against the Soviet Union. It was a techno-nationalist show of prowess. Today, post–September 11 and, equally or more important, with the ongoing war in Iraq, the United States needs to again recognize and embrace the leadership opportunity offered by manned space exploration, but this time based on cooperation, not competition. Leading an international, inclusive expedition from earth allows the United States to counter its unilateralist, militarist image, which has prevailed due to both the Iraq war and U.S. moves toward space weaponization. Such a choice would go a long way toward rebuilding American soft power by positively leading the world on a global endeavor to step into space together, for exploration, development, and applications useful on earth. It is the ultimate positive "big event" strategic communication message of leadership. From the global participants' side, taking part in a grand space program does more than just help countries construct technology and create industries; it builds dreams and generates pride. Working cooperatively with other countries on a space venture would also alleviate fears about U.S. intentions to monopolize space. The United States has demonstrated its military ability to make others bend to its will. Now it must work at not having to use that ability. Soft power is essential to build a stable, peaceful world in which the human security needs of all are met.

The Bush Space Vision was intended to be bold; it hoped to show the innovative and daring spirit of the United States, which is admirable. However, lessons learned from the past seem to have been ignored, and extrapolations forward or considerations of new models seem not to have been made. The capabilities and motivations of other countries have also not been considered. Finally, the United States has an unrealistic budget for its space ambitions. All this has left the United States in a position where failure, as much as one hopes otherwise, is not just possible but probable. It is increasingly unlikely that the United States will be able to finish the ISS within the original timetable[22]—or meaningfully move forward with the new transportation vehicle necessary to realize the new space vision. This is especially true given the continuing problems with the shuttle. The United States needs to regroup, and there are several obvious parameters for consideration:

Do not abrogate commitments on one international program to begin another one. The current web of mismatched commitments and time lines needs to be unraveled. The United States committed to completing the ISS with international partners. Abrogating one commitment while trying to interest the same partners into entering a new commitment is incongruent at best. President Bush, in his January 2004 vision announcement, stated that "our first goal is to complete the International Space Station. We will finish what we have started, we will meet our obligations to our 15 international partners." The United States needs to follow through.

Establish realistic timelines. If the ISS is to be finished, and it should be, then time lines have to be rewritten. Since the vision in its current form is by all accounts unworkable, there appears to be time to finish the station while developing a realistic way forward on what is here referred to as the Space Exploration Partnership (SEP). It is more important to do it right than to do it quickly. The "rush to failure" approach that the missile defense program followed is not to be repeated.

Clearly, a new space transportation vehicle is imperative to any new program. Whether or not the crew exploration vehicle is the right one or not is debatable. Every attempt should be made to develop a new vehicle, whatever it may be, but the timetable for its development should not rely on or compete for funds allocated for the shuttle or the ISS. Funding competition and unrealistic timelines create potential safety conflicts of interest that must be avoided. They are a recipe for failure.

Develop the SEP as a strategic program, not a science program. While the annual budget process in the United States inherently makes long-term budget commitments on the part of the United States impossible, political commitments do vary significantly, according to program purpose. Apollo was a strategic program and was consequently funded to completion because its goals were considered important to American security. The SEP must be developed along the same rationale. The purpose of the SEP must be understood by policy makers and conveyed to the public as restoring the U.S. global leadership image, not only exploring space. The SEP becomes a program of positive patriotism, America as the "shining city on the hill," to contest the images of negative, chest-thumping patriotism and militarism that much of the world sees as pervasive. This is imperative for U.S. efforts to build a more stable world, free from terrorism.

Develop the SEP inclusively rather than exclusively. All interested countries must be included. The United States has had more success in shaping the space activities of other countries through cooperation than competition. Cooperation in space with China does not excuse the Communist regime from its commitment to Communism or its abysmal record on human rights. But precisely because China is an authoritarian state at the crossroads of its political development, the United States, as the world's leading democracy, must help shape China's aspirations in space toward peaceful and cooperative ends, rather than see them turned toward more threatening ideological or military goals. At a press briefing on April 20, 2006, Dennis Wilder, acting senior director of the National Security Council, talked about President Bush's meetings with President Hu Jintao as related to space:

> The President, in the area of trying to deepen the relationship between our two societies and our two cultures, offered to send the NASA Administrator to China to begin to talk about lunar exploration with the Chinese, to talk about some of the things we need to do in space—for example, debris avoidance and other subjects. There are some things that the Chinese also have in terms of sensor technologies and information that we are interested in, in terms of global climate and other issues. So the NASA Administrator will probably go to China later on this year to begin to consult on the subject of space exploration and where we might

> have common interests and where we might begin to work together as the two nations on the Earth with the most ambitious space programs in the 21st century at this point.[23]

The visit by NASA administrator Mike Griffin to China offered a timely opportunity to begin a new and improved space relationship with China. Whether space cooperation between the United States and China is undertaken as part of a "grand bargain" or initiated incrementally, the benefits such cooperation offers the United States far outweigh the potential risks. The key to success will be if functionaries can and will follow through with the president's apparent new strategic guidance. Griffin is a known supporter of international cooperation and has long experience in working with partners in ways beneficial to all parties. While a change of policy is not up to him alone, he is well suited to move an initiative forward.

There will also be those in both China and the United States who do not want to see relations improve. Every so often, opponents of greater cooperation in space rediscover the "Chinese laser threat" to American satellites. It is unlikely a coincidence, for example, that on the first day of Griffin's visit to China, news stories appeared—citing the usual unnamed experts—that decried supposed recent Chinese attempts to blind U.S. satellites with lasers.

Such stories are to some extent true, but they are also unremarkable. The Chinese have on occasion "dazzled" (rather than permanently blinded) American satellites with ground-based lasers, both to test Chinese capabilities and to remind the Americans that dominating space will not be easy. No doubt, it reflects as well a certain amount of chest-thumping by those in China who are not keen to cooperate with the nation they regard as their chief enemy in the world. Still, no American satellites have actually been damaged or even interfered with, despite nearly a decade of this sort of orbital mischief. But that does not stop the "laser threat" story from being dragged out at opportune moments—along with other urban legends, such as the infamous "Chinese killer mini-sats" story—as proof that cooperation with China is a dangerous idea.

Link the ISS and the SEP. If the SEP includes both Russia and China, there might be an opportunity to make a package-deal arrangement with either or both for transportation to the ISS while it is being finished as part of their contribution to the new program. With continuing issues with the shuttle, the imperative is further heightened. Linking the ISS and the SEP rather than jettisoning one to start the other alleviates the concerns of the ISS partners, maximizes U.S. investments already made, and facilitates a transportation deal among partners from which all benefit.

Bring in partners early. While the United States will, by virtue of intent, stature, resources, and technology, lead the SEP, the partnership will be strongest if all involved feel they have a voice in determining program goals. The program should begin with a space summit to bring in potential partners early and give them such a voice. The two NASA exploration strategy workshops held in 2004 and 2006 have been intended as a step in that direction. The fog and friction surrounding implementation plans and U.S. commitment to completing the ISS have left potential partners hesitant, however. Additionally, the United States must be prepared to accept partners, not participants. Some countries could be partners, giving them a voice in decision making, depending on their contributions, or what works best politically.[24]

Cooperative programs can be accomplished with partners providing unique hardware to minimize the risk of technology transfer, but the entire area of technology transfer and export control is still sorely in need of reexamination as part of a comprehensive space policy. For the United States to lead in space, its aerospace industry must be healthy and globally competitive. Globalization should benefit the industry, not put it at risk.

Maintaining Commercial Space Leadership

Truth is often stranger than fiction. In February 2005, DirecTV Group agreed to pay $5 million in fines for having violated export-control regulations incurred in connection with providing China with technology "to enhance secure satellite communications." DirecTV's subsidiary, Hughes Network Systems, sold voice and data transmission equipment to China, India, South Korea, Turkey, and South Africa between 1993 and 2003. Apparently, the company modified the equipment and provided the always nebulous "services" related to the sales without appropriate—and who knows what that means—government approval. Encryption, increasingly an integral part of commercial data transmission requirements to protect against identity theft, can also be of value to the military. The U.S. government was concerned that equipment sold to China with encryption capabilities, regardless of intended purposes, could find its way to the Chinese military. But at the same time that communications technology remains on the "sensitive" list, and American companies continue to be beaten with the export-control club if they risk selling it abroad, the United States is considering relaxing export controls on missile technology, which it is spending billions to defend against.

The United States sees a threat to missile proliferation, so it builds a missile defense system to defend against those missiles. Admirably, to show that it is not building the system either to create Fortress America or to have both the Sword and the Shield against the rest of the world, and to get access to needed global ground stations, the United States seeks international partners. For the partners to participate, however, missile technology must be transferred, which would violate export controls. So the government is considering relaxing the controls imposed by the Missile Technology Control Regime (MTCR), which the United States has long supported and been a member of, and encouraged other countries—specifically China—to join and support as well.

If the United States revises the MTCR export controls, technology would initially be transferred to only a handful of countries in partnership with the United States on missile defense. However, a second tier of countries might be considered as well—those considered U.S. friends and allies. But which are they? Those on the list for consideration include Israel, Taiwan, South Korea, and even Pakistan and Egypt.[25] How to protect against subsequent transfers has long been the rationale for what many have considered an overly restrictive export-control policy on such technologies as voice and data transmission equipment. Even the closest and most reliable U.S. allies have had technology slip through their fingers. For missiles, however, apparently the risks are considered worth taking. Whether missile defense is worth the risk or not—a question well worth asking—or whether it ends up creating a bigger missile problem than we started with, abrogating MTCR controls would certainly exacerbate an already irrational export-control environment in the United States. It also supports the view that the United States adheres to the rule of law and international regimes only when convenient, a view that must be dispelled.

Untangling the export-control process remains an American industry priority, and should be a government priority as well, though it is not. To raise the issue's importance for government officials, the Aerospace Industries Association, one of the largest aerospace industry trade associations, hired a new vice president of international affairs in June 2005. The AIA announced that Rear Admiral Craig Steidle, formerly in charge of executing the Bush Space Vision for NASA, would be spearheading industry efforts "to promote government policies that support exports, avoid protectionism, and foster fair principles of international trade."[26] Whether extricating Alice from Licenseland will prove any easier than returning the United States to the moon, and reaching Mars and beyond, remains to be seen.

Reforming its export-control regulations and processes is in the national interest of the United States, to protect a critical national industrial sector, create a nonproliferation plan that makes sense and is globally viable, and address global concerns that the United States is deliberately denying technologies to developing countries so that it can exploit them economically, politically, and culturally. The current system negatively affects all those goals and is the worst of all possible worlds. Nevertheless, the system continues to have its supporters.

According to Jon Howland of the Jewish Institute for National Security Affairs, "If the United States is to remain the leader in space, law makers and military planners within the U.S. government need to appreciate the inherent dangers involved in the continued proliferation of dual-use technology under the guise of commercialization."[27] Certainly the dilemmas of dual-use technology are appreciated. But does that mean that the United States should be the only country allowed to own or even have access to the benefits of space technology? That would be the equivalent of twenty-first-century colonialism. Since a U.S. monopoly is, first, impossible because the technology has been globalized and, second, not conducive to fostering the type of global cooperation needed to fight the global war on terrorism and address other transnational issues, even the appearance or perception of support for that option is one to avoid, both directly and as an unintended consequence of other actions.

Export-control reform does not mean choosing between commerce and national security, nor does it always mean loosening controls. Export regulations are intended to strengthen national security by providing the United States with some rational "control" over technology. Since the end of the Cold War, however, and especially since the Cox Committee in the late 1990s, just the opposite has occurred. By trying to stop what it ultimately does not really control, the United States increasingly ends up with 100 percent control of nothing. This dizzying dance must be stopped.

While eliminating conflicting legislative mandates about export controls is imperative, as a congressional responsibility, it is in all likelihood unlikely to occur quickly or easily. Attempts that have been under way for more than ten years seem to bear out that conundrum. In the meantime, however, other actions can still be very useful:

Set achievable export-control goals. This will require a restructuring of the munitions list. The nearly 10,000 items on it include many that neither are sensitive nor contain or involve unique technologies, some of which are available commercially in the United States. Some very tough decisions have to be made. If an item is commercially available in the United States, should it still be subject to export-control regulations? If an item is available from another country, does it make sense for the United States to prohibit

American manufacturers from selling it? Is the United States better off in the long run regulating technology proliferation rather than trying to stop it—that is, allowing a less than newest version of the technology be made available for export? Once the initial "scrub" of the list is done, the list should then be reviewed regularly and transparently.

The United States must also determine which countries it is trying to keep technology from. Clearly, the same set of rules should not apply to Canada and Great Britain as would apply to China or countries in the Middle East. For some countries, blanket exemptions for entire categories of technology might be granted, though the number of countries receiving such exemptions should be limited, if only because technology can outlast friendships. Equipment that the United States provided to Afghanistan and Iraq earlier was used against U.S. troops in 2001 and 2002. Do we really now want to send missile technology to Pakistan and Egypt?

Considerations of third-party transfers also need to be considered, which will most likely involve multilateral discussions toward building an agreed export-control framework for all. Admittedly, agreement will likely not be as defined and strict as the United States would want. But perhaps agreement can be reached on what technology should be absolutely prohibited for export sale, a useful step forward. That model is useful in many areas.

Clarify the export-control rules. Clearly defining what is allowed and what is not is critical in each proposed reform, and for the process in general. The aerospace industry is important to U.S. national security, and it must not be put in a disadvantaged position when making a transaction with a foreign country. Companies should not have to guess what is allowable and then find themselves in the wrong later.

A first step in this direction would be to clarify what items specifically are under the purview of the Department of Commerce for licensing and the rationale behind the selection of the items. Knowing what is on the list, agreed to by Commerce, State, and Defense, will expedite the process. Knowing the rationale might also help the aerospace industry avoid some potential potholes by better understanding why one item is considered "sensitive" and another is not.

Simplify the export-control rules. Because of the dilemmas of dual-use technology, another suggested approach has been to move from a transactional approach to licensing to one of a more process-based approach. The current system potentially requires a separate license or technology assistance agreement for each transaction, whether a meeting, sending a part from the manufacturer to the buyer, returning a defective part, or other such activity. In other words, unless the government specifically says a company may do something, it may not do anything. This is incredibility inefficient and costly to both the buyer and the seller. Parameters should be given that approve limits of acceptable activity, anything beyond which requires a license or TAA. This incorporates techniques already used in quality-control processes, which say that it is acceptable to proceed up to a certain boundary. There are still a multitude of issues that remain to be worked out, and on which reasonable people may disagree, including many definitional issues. But such a system would give the aerospace industry some breathing space and negate some of the lure of European ITAR-free satellite advertisements. Foreign firms and governments want U.S. aerospace products, but not at the risk of betting the company on a nebulous U.S. export-control regime that decides their fate behind closed doors, sometimes over prolonged periods, based on seemingly arbitrary and ambiguous rules.

Clarify organizational responsibility. Make it clear who is responsible for executing the rules and hold those people responsible for timely, clear answers. Sellers and buyers should be held accountable for abiding by export-control regulations; those responsible in the government for implementing the regulations should be held accountable as well. While the vast majority of license applications are eventually approved, some are not, and some take much longer than the median time period of thirty days. Use of median statistics rather than averages is a deception in itself, since median figures mitigate the statistical impact of those that take ten times the average. Further, if the license sits in processing for 120 days and is then returned with no action taken, it disappears from the statistics. During processing, the applicant can go online and see the status of its license, but can do nothing about it if the application is stalled. Not being able to find out why an application is being delayed, and not having the opportunity to explain or fix whatever issue is holding it up—or even speak with someone on the telephone about the issue—is unacceptable.

Transparency and efficiency. A transparent and efficient administrative process for license examination and issuance must be in place. That requires adequate funding to deal with the number of licenses to be processed. While that number would be substantially reduced if the munitions list was revised or a process-based approach adopted, well-trained personnel in sufficient numbers are still needed. Then, the process itself must be streamlined and made transparent, to avoid decisions being made by unnamed, faceless people behind closed doors on an apparently arbitrary basis. In some cases, streamlining involves making sure that all the involved government offices can electronically talk to one another and share files. In others, it involves making sure that delays are neither inadvertently nor intentionally imposed due to bureaucratic or governmental turf battles. Finally, toward allowing applicants to function better in the business world while allowing government concerns to be addressed, a government–industry consultation process should be established as part of the process, to make export controls logical and useful to all.

The Pentagon's *Military Power of the People's Republic of China*, an annual report to Congress, usually released in May of every year, was finally released on July 20, 2005.[28] The grapevine story on the delay was that there were disagreements between Pentagon hardliners and those with more reasoned views at the CIA and State Department on the tone of the report. Judging from the space section, the latter seemed to have won out. Language about Chinese parasite satellites and speculation about Chinese intent was gone, replaced by toned-down rhetoric concerning what "China is studying" and what "China is interested in." Apparently, reason prevailed.

Space is a subset of foreign policy. For the space ambitions of the United States to prevail without an unintended and counterproductive clash with other countries, the connectivity between activities in the manned, military, and commercial arenas must be not only recognized, but considered and factored into decision making. If the United States expands its military space capabilities and continues to thwart the spread of space technology, it can be expected that other countries will expand their efforts to obtain the technology they feel is needed to protect their sovereignty. If the United States intends to retain space superiority through dominance, a healthy aerospace industry is critical. If the United States abrogates its leadership position in manned spaceflight, China will be willing to assume the position by default, with all the techno-nationalists perceptions that accrue with that position. If the United States needs global cooperation to fight the global war on terrorism, cooperation in manned space provides an opportunity to foster cooperation, gather soft power, and counter perceptions that U.S. military space activity seeks to shut other countries out of the heavens. Such a space program, however, requires the United States to recognize the inherent links between its space efforts and its foreign policy. Such links are rarely acknowledged, let alone factored into decisions.

Iranian writer Azar Nafisi succinctly summarizes the U.S. worldview compared with those of other countries: "We in ancient countries have our past—we obsess over the past. They, the Americans, have a dream: they feel nostalgia about the promise of the future."[29] America is a forward-looking country, focusing on a positive future. Space has always been a part of America's future.

It is imperative that it remain so. If we approach it correctly, perhaps we can even convince those obsessed with the past that a balance between past and future can be achieved, to the benefit of all.

The United States has the opportunity to use space to its strategic advantage, as it did earlier with Apollo. Political will and public support are the keys to success. Both require reasoned and informed discussion and debate on all the issues. A clash of ambitions in space is no more inevitable than is the weaponization of space. There is still time to change direction and deescalate the security dilemma. Without a change in direction, however, the United States could increasingly find itself the odd man out in cooperative space projects or leading projects with minimal international commitment—relying on an increasingly strained aerospace industry and in a constant race with itself to create bigger, better, and always more expensive military space technology. Space is a frontier of seemingly unlimited horizons, yet one that has been parceled into the worst kind of policy stovepipes. It is time that the United States look at the big picture and, based on that, determine a best course of future action. Space security can be achieved in more than one way, and it is imperative that we explore them all.

Notes

Preface

1. George Abbey and Neal Lane, *United States Space Policy: Challenges and Opportunities* (Cambridge, Mass.: American Academy of Arts and Sciences, 2005), 11–15.
2. Other political clashes have been suggested. See, for example, Samuel P. Huntington, "The Clash of Civilizations," *Foreign Affairs* 72, no. 3 (1993): 22–28; and Stanley Hoffman, "Clash of Globalizations," *Foreign Affairs* 81, no. 4 (2002): 104–115.

1. A Clash of Ambitions

1. *Yuhangyuan* is Chinese for "space navigator" and is used in official Chinese media. *Taikonaut* is derived from *taikong*, the Chinese word for "space," and has been used primarily outside China.
2. *Shenzhou* is translated many ways, "divine vessel" and "magic boat" being the most common.
3. Alexander Rose, "Bush's Space Vision," *National Review*, February 28, 2005.
4. Tim Weiner, "Air Force Seeks Bush's Approval for Space Weapons Program," *New York Times*, May 18, 2006.
5. Peter Brookes, "Militarizing Space," *New York Post*, June 7, 2005.
6. Michael Krepon and Christopher Clary, *Space Assurance or Space Dominance? The Case Against Weaponizing Space* (Washington, D.C.: Henry L. Stimson Center, 2003).
7. James Oberg, "Hyperventilating over 'Space Weapons,'" *USA Today*, June 14, 2005.
8. Oberg, "Hyperventilating over 'Space Weapons.'"

9. Karl Mueller, "Is the Weaponization of Space Inevitable?" (paper presented at the International Studies Association Annual Conference, New Orleans, March 17, 2002).
10. Ed Crane, "Memo to Karl Rove," *Wall Street Journal*, April 24, 2005.
11. Carl Sagan, *The Demon-Haunted World: Science as a Candle in the Dark* (New York: Random House, 1997); Charles P. Pierce, "Greetings from Idiot America," *Esquire*, November 2005, 181–185.
12. Weiner, "Air Force Seeks Bush's Approval for Space Arms."
13. Walter Russell Mead, *Special Providence* (New York: Routledge, 2002), 98.
14. Former president Ronald Reagan drew that reference from a 1630 sermon by John Winthrop, which was based on the same "city upon a hill" scripture from Matthew 5:14.
15. Robert Jervis, "Cooperation Under the Security Dilemma," *World Politics* 30, no. 2 (1978): 167–214.
16. Barry R. Posen, "The Security Dilemma and Ethnic Conflict," *Survival* 35, no. 1 (1993): 27–57; Thomas J. Christensen, "China, the U.S.–Japan Alliance, and the Security Dilemma in East Asia," *International Security* 23, no. 4 (1999): 49–80, and "The Contemporary Security Dilemma: Deterring a Taiwan Conflict," *Washington Quarterly* 25, no. 2 (2002): 7–22; William Rose, "The Security Dilemma and Ethnic Conflict: Some New Hypotheses," *Security Studies* 9, no. 4 (2000): 1–54; Robert S. Ross, "Beijing as a Conservative Power," *Foreign Affairs* 76, no. 2 (1997): 33–45.
17. There is a substantial body of literature on the influence of perception on international relations generally, and on how policy makers develop their perceptions specifically. See, for example, Alexander Wendt, "Anarchy Is What States Make of It: The Social Construction of Power Politics," *International Security* 16, no. 4 (1992): 391–425; and Robert Jervis, *Perception and Misperception in International Politics* (Princeton, N.J.: Princeton University Press, 1976).
18. Jervis, "Cooperation Under the Security Dilemma," 201.
19 Though John F. Kennedy's "go to the moon" speech is popularly cited as the beginning of the Apollo program, NASA initiated the Apollo program in July 1960 as a follow-up to the Mercury program.
20. Jervis, "Cooperation Under the Security Dilemma," 200.
21. Stefano Silvestri, "Space and Security Policy in Europe," executive summary, occasional paper, no. 48, Institute for Security Studies, European Union, Paris, December 2003, 9.
22. Department of Defense Directive 3100.10, July 1999; Joint Chiefs of Staff, Department of Defense, *Joint Doctrine for Space Operations*, Joint Publication 3-14 (Washington, D.C.: Department of Defense, 2002), GL-2, available at http://www.dtic.mil/doctrine/jpoperationsseriespubs.htm (accessed June 20, 2006).
23. Pew Global Attitudes Project, *U.S. Image Up Slightly, But Still Negative: American Character Gets Mixed Reviews* (Washington, D.C.: Pew Global Attitudes Project, 2005), cited in Andrew Kohurt and Carolyn Funk, "Poll: In Wake of Iraq War, Allies Prefer China to U.S.," June 24, 2005, available at www.pewglobal.org/reports/pdf/247.pdf (accessed August 13, 2006).
24. "Putin Welcomes China as New Member of Space Powers Club," Global News Wire, reported in BBC Monitoring International Reports, October 16, 2003.

25. European Space Agency, "ESA Director General Salutes China's First Human Space Flight," press release, October 15, 2003.
26. "Blair's New Europe, or Europe as Counterbalance to the USA," *Die Welt*, February 15, 2003, reported in BBC Monitoring Europe, February 17, 2003.
27. The periods from the mid-nineteenth century to the late 1920s, and from 1945 to 1989, are considered earlier iterations of globalization because of the high degree of international connectivity that occurred.
28. Ted Bridis, "Bush Prepares for Possible Shutdown of GPS Network in National Crisis," Associated Press, December 16, 2004.
29. Scott Pace, Gerald P. Frost, Irving Lachow, Dave Frelinger, Donna Fossum, Don Wassem, and Monica M. Pinto, *The Global Positioning System: Assessing National Policies* (Santa Monica, Calif.: RAND Corporation, 1995), 117; Communication for the Commission, "Galileo—Involving Europe in a New Generation of Satellite Navigation Services," February 10, 1999, available at http://europa.eu/scadplus/leg/en/lvb/l24205.htm (accessed August 13, 2006).
30. Directorate General Energy and Transport, European Commission, Galileo, "The European Project on Radio Navigation by Satellite," press release, March 26, 2002.
31. Russia maintains the Glonass navigation satellite system, though recent financial straits have left it virtually moribund. China has a system as well, called Beidou, with limited regional capabilities.
32. The president's vision, documented as "A Renewed Spirit of Discovery, the President's Vision for U.S. Space Exploration" (available at www.nasa.gov/pdf/55583main_vision_space_exploration2.pdf), was followed by another NASA online document, "The Vision for Space Exploration" (http://www.nasa.gov/mission_pages/exploration/main/index.html), that provides NASA strategy and guiding principles.
33. NASA has field centers in California, Florida, Texas, Mississippi, Ohio, Maryland, Virginia, and Alabama.
34. Darren K. Carlson, "To Infinity and Beyond on a Budget," Gallup Poll Briefing, August 17, 2004.
35. However, the last three scheduled missions were canceled due to lack of interest and, subsequently, funding.
36. Carlson, "To Infinity and Beyond on a Budget."
37. Guy Gugliotta, "NASA Unveils $104 Billion Plan to Return to the Moon by 2018," *Washington Post*, September 20, 2005.
38. *Star Wars*, dir. George Lucas, Lucasfilm, San Francisco, 1977.
39. Craig Covault, "Desert Storm Reinforces Military Space Directions," *Aviation Week & Space Technology*, April 8, 1991, 42.
40. See, for example, Air Force, Assessment and Analysis Division, *Operation Iraqi Freedom—By the Numbers* (Washington, D.C.: United States Air Force, 2003).
41. Christopher J. Bowie, Robert P. Haffa Jr., and Robert E. Mullins, *Future War: What Trends in America's Post-Cold War Military Tell Us About Early 21st Century Warfare* (Washington, D.C.: Northrop Grumman Analysis Center Papers, 2003), 46.
42. Air Force, *Operation Iraqi Freedom*.

43. Ann Imse, "U.S. Deploys Weapon to Attack Satellites," *Rocky Mountain News*, October 1, 2004.
44. Elaine M. Grossman and Keith J. Costa, "Small, Experimental Satellite May Offer More than Meets the Eye," *Inside the Pentagon*, December 4, 2003.
45. U.S. Space Command, *Vision for 2020* (Peterson Air Force Base, Colo.: U.S. Space Command, 1997), 5; Jim Garamone, "Joint Vision-2020 Emphasizes Full-Spectrum Dominance," Armed Forces Press Services, available at http://www.defenselink.mil/news/Jun2000/n06022000_20006025.html (accessed July 6, 2006).
46. Donald Rumsfeld, chair, "Report of the Commission to Assess United States National Security Space Management and Organization," January 11, 2001, viii, ix, available at http://www.defenselink.mil/pubs/space20010111.html (accessed July 6, 2006). The phrase "Space Pearl Harbor" was used seven times in the report.
47. U.S. Air Force Deputy Chief of Staff for Plans and Programs, *The U.S. Air Force Transformation Flight Plan* (Washington, D.C.: U.S. Air Force Future Concepts and Transformation Division, 2003), available at www.af.mil/library/posture/AF_TRANS_FLIGHT_PLAN-2003.pdf (accessed June 26, 2006).
48. Eric Adams, "Is This What War Will Come To?" *Popular Science*, June 2004. The artist's rendering was by John Macneill.
49. Thomas M. Nichols, "Anarchy, Order, and the New Age of Prevention," *World Policy Journal* 22, no. 3 (2005): 1–24.
50. Jervis, "Cooperation Under the Security Dilemma," 173
51. Neelam Mathews and Frank Morring Jr., "Dare to Dream; Indian President Draws on a Career in Space, Missile Work to Design His Nation's Future," *Aviation Week & Space Technology*, November 22, 2004, 51.
52. Walter Russell Mead, *Power, Terror, Peace, and War* (New York: Knopf, 2004).
53. Joseph Nye, *The Paradox of American Power: Why the World's Only Superpower Can't Go It Alone* (New York: Oxford University Press, 2002).
54. "Shallow Machiavellis," Newsweek International, *Newsweek Issues 2004: Special Report—Power—Who's Got It Now*, available at http://www.msnbc.msn.com/id/3606144/site/newsweek/ (accessed June 20, 2006).
55. Union of Concerned Scientists, "Countermeasures: A Technical Evaluation of the Operational Effectiveness of the Planned US National Missile Defense System," report summary, April 2000, available at http://www.ucsusa.org/global_security/missile_defense/countermeasures.html (accessed August 13, 2006); Andrew Erickson, "Chinese BMD Countermeasures: Breaching America's Great Wall in Space?" in *Newport Paper*, no. 22, ed. Lyle J. Goldstein (Newport, R.I.: Naval War College Press, 2005).
56. Jervis, "Cooperation Under the Security Dilemma," 182–183.

2. The Conundrum of Dual-Use Technology

1. Department of Defense, "Dual Use Science and Technology," available at http://www.acq.osd.mil/ott/dust/ (accessed June 20, 2006).

2. The source of the statistics appears in International Telecommunications Union, "Key Global Telecom Indicators for the World Telecommunications Services Sector," available at http://www.itu.int/ITU-D/ict/statistics/at_glance/KeyTelecom99.html (accessed June 20, 2006).
3. DUST homepage, available at http://www.acq.osd.mil/ott/dust/index.html (accessed July 10, 2006).
4. Air Force, Assessment and Analysis Division, *Operation Iraqi Freedom—By the Numbers* (Washington, D.C.: United States Air Force, 2003), 12.
5. Office of the Under Secretary of Defense, *MCTL*, Section 17—Space Systems Technology (Washington, D.C.: Department of Defense, 1996), 17-1.
6. Institute for National Strategic Studies, National Defense University, *Strategic Assessment 1995: U.S. Security Challenges in Transition* (Washington, D.C.: Government Printing Office, 1995), 151.
7. It was later purchased by Space Imaging.
8. In September 2004, Space Imaging sold its Federal Civil/Commercial Solutions business unit to the Denver-based Geo360 Corporation. The group provides mapping and geographic information system (GIS) services to federal, state, and local government customers.
9. Determining overall capabilities must go beyond resolution. Other factors, such as the quality of the imagery, maneuverability of the satellite, life cycle of the system, and analysis of the product, are key. Satellite maneuverability in part determines how near to real-time an image can be obtained. In some instances, the time required to position the satellite can detract from the value of the imagery to users or can limit the number of different sites that can be viewed. Different satellites also have different positioning speeds.
10. Katie Hafner and Saritha Rai, "Google Offers a Bird's-Eye View, and Some Governments Tremble," *New York Times*, December 20, 2005, 1.
11. Office of the Under Secretary of Defense, *MCTL*, 17-1.
12. International Telecommunications Union, "Key Global Telecom Indicators for the World Telecommunications Services Sector"; see also Economist Intelligence Unit, *Telecoms and Technology Forecast China May 2004* (London: Economist Intelligence Unit, 2004), 2.
13. Theresa Hitchens, *Future Security in Space: Charting a Cooperative Course* (Washington, D.C.: Center for Defense Information, 2004), 39, 41.
14. There are other orbits as well, including polar orbits, which are especially useful for weather satellites, and a specialized orbit called Molniya, used mostly by Russia because of its extended coverage of northern latitudes.
15. Hitchens, *Future Security in Space*, 40.
16. For a history on Landsat, see Pamela Mack, *Viewing the Earth: The Social Construction of the Landsat Satellite System* (Cambridge, Mass.: MIT Press, 1990).
17. John C. Baker, Kevin M. O'Connell, and Ray A. Williamson, *Commercial Observation Satellites* (Santa Monica, Calif.: RAND Corporation, 2001).
18. G. Teo, "Post Officers Use High-Tech Tools to Catch Oil Slick Culprit," *Straits Times*, January 19, 1997.

19. Michel Bourbonneire, "Law of Armed Conflict (LOAC) and the Neutralisation of Satellites, or *Ius in Bello* Satellites," *Journal of Conflict and Security Law* 9, no. 1 (2004): 43–69.
20. Dwayne A. Day, "A Look at . . . Spy Satellites and Hollywood," *Washington Post*, July 2, 2000, B3.
21. Scott Pace, Gerald P. Frost, Irving Lachow, Dave Frelinger, Donna Fossum, Don Wassem, and Monica M. Pinto, *The Global Positioning System: Assessing National Policies* (Santa Monica, Calif.: RAND Corporation, 1995), 243.
22. Pace et al., *Global Positioning System*, xx.
23. Civilian users can and do get much greater accuracy on occasion.
24. "Garmin Guys Garner Glory," *GPS World*, November 2003.
25. Defense Update, "Joint Direct Attack Munitions (JDAM)," available at http://www.defense-update.com/products/j/jdam.htm (accessed June 20, 2006); Air Force, *Operation Iraqi Freedom*.
26. The need to carefully track civilian aircraft came to light after the shooting down of Korean Airlines flight 007 in 1983, after it strayed off course and into Soviet airspace.
27. National Security Council, "Discussion at the 339th Meeting of the National Security Council, Thursday, October 10, 1957," October 11, 1957, NSC Series, box 9, Eisenhower Papers, 1953–1961 (Ann Whitman File), Dwight D. Eisenhower Library, Abilene, Kansas, available at history.nasa.gov/sputnik/oct57.html (accessed July 28, 2006).
28. "U.S. Scientific Satellite Program," in John Logsdon, ed., *Exploring the Unknown: Selected Documents in the History of the U.S. Civil Space Program*, vol. 2, *External Relations*, NASA History Series, SP-4407 (Washington, D.C.: NASA History Office, 1996), 241.
29. Walter McDougall, . . . *The Heavens and the Earth: A Political History of the Space Age* (Baltimore: Johns Hopkins University Press, 1997), 123.
30. Georgi Grechko, conversation with author, 1991.
31. George R. Price, "Arguing the Case for Being Panicky," *Life*, November 18, 1957, 125, 6, 128.
32. "The Feat that Shook the World," *Life*, October 21, 1957, 23.
33. One perspective was that improving commercial launch technology could lead to improving the Chinese ability to target maneuverable reentry vehicles (MARVs) and avoid operational missile defenses. This view assumes nuclear weapons to be warfighting tools rather than deterrents. As a warfighting tool, whether on target or not, it is likely that if China launched a nuclear weapon at the United States, the U.S. reaction would involve the use of nuclear weapons as well.
34. Much of the technology for the Delta came from the Thor IRBM program.
35. Roger B. Handberg and Joan Johnson-Freese, *The Prestige Trap* (Dubuque, Iowa: Kendall-Hunt, 1994), 170–171.
36. Vernon Loeb, "U.S. Satellite Sales Lag Since Regulatory Shift," *Washington Post*, March 28, 2000, A21.
37. International Traffic in Arms Regulations, July 9, 2001, sec. 129.2, 22 *CFR* 129.2.

3. From Apollo to Where?

1. Marcia S. Smith, *U.S. Space Programs: Civilian, Military, and Commercial*, issue brief IB92011, prepared by Congressional Research Service, updated April 22, 2003, summary page and graph.
2. Francis Fukuyama talks about the dilemma created in foreign affairs by this American tendency to lose interest in issues: "Americans are not, at heart, an imperial people. Even benevolent hegemons sometimes have to act ruthlessly, and they need a staying power that does not come easily to people who are reasonably content with their own lives and society" ("After Neoconservatism," *New York Times Magazine*, February 19, 2006).
3. Gwyneth K. Shaw, "Race Doesn't Reflect NASA, Exploration," *Orlando Sentinel*, October 12, 2004.
4. Otto Kreisher, "Aerospace Industry Seeks Candidate Support," Copley News Service, October 4, 2004.
5. John M. Logsdon, "A Consistent Theme: Support for Space Exploration," in *The Case for Space Exploration* (Colorado Springs: Space Foundation, 2006), 26–29.
6. A 2004 Gallup opinion poll and report (available at www.spacecoalition.com) indicate public support for the manned space program.
7. Al Neuharth, "Next Huge Space Shock: China," *Florida Today*, May 14, 2003; "This Time Around, U.S. Trails in Race to Moon," *Providence Journal*, November 21, 2004, A18.
8. Interview with T. Keith Glennan, in *NASA 25th Anniversary Video History Series*, episode 1, *The Birth of NASA*, prod. NASA, 1984 (NASA-TM-110819).
9. "Institutional failure" was considered to be a cause of both disasters. See Marcia S. Smith, *NASA's Space Shuttle Columbia: Synopsis of the Report of the Columbia Accident Investigation Board*, report RS21606, CRS-4, prepared by Congressional Research Service, September 2, 2003.
10. Congressional Budget Office, "A Budgetary Analysis of NASA's New Vision for Space Exploration," September 2004, available at http://www.cbo.gov/showdoc.cfm?index=5772&sequence=0 (accessed August 13, 2006), and "Alternative Costs for Boost-Phase Missile Defense," sec. 7 of 9, July 2004, available at http://www.cbo.gov/showdoc.cfm?index=5679&sequence=1.
11. Field centers are the Ames Space Center, California; Dryden Flight Research Center, California; Glenn Research Center, Ohio; Goddard Space Flight Center, Maryland; Jet Propulsion Laboratory, California; Johnson Space Center, Texas; Kennedy Space Center, Florida; Langley Research Center, Virginia; Marshall Spaceflight Center, Alabama; Stennis Space Center, Mississippi.
12. Sean Holton, "Whitten's Influence Kept NASA Factory, Money in His District," *Orlando Sentinel*, October 11, 1992, A12.
13. Quoted in Brian Berger, "Expert Panel Tells Congress: NASA Needs New Direction," Space.com, October 16, 2003, available at http://www.space.com/missionlaunches/nasa_hearing_031016.html.

14. For an overview of international cooperation and competition historically, see Joan Johnson-Freese, *Changing Patterns of International Cooperation in Space* (Malabar, Fla.: Orbit Books, 1990).
15. 1958 Space Act. Pub.L. 85-568, July 29, 1958, 72 Stat. 426, 42 U.S.C. § 2451.
16. NASA, Advisory Council, Task Force on International Relations in Space, *International Space Policy for the 1990's and Beyond*, October 12, 1987, 18.
17. A one-day seminar on June 6, 2005, sponsored by the American Astronautical Society and the American Institute of Aeronautics and Astronautics, specifically looked at case studies in international cooperation and setting out lessons learned. See http://www.astronautical.org/index.php?option=com_content&task =view&id = 22&Itemid=1 (accessed July 31, 2006).
18. American Institute of Aeronautics and Astronautics, *From Challenges to Solutions* (report of the seventh International Space Cooperation Workshop, Anchorage, Alaska, May 2–6, 2004). See also Peggy Finarelli and Ian Pryke, "Optimizing Space Exploration Through International Cooperation" (paper presented following the seventh International Space Cooperation Workshop, Anchorage, Alaska, May 2–6, 2004).
19. Joan Johnson-Freese, "The International Solar Polar Mission," *Space Policy* 3, no. 1 (1987): 24–37.
20. Marcia S. Smith, testimony before House Science Committee, *NASA's Space Station Program: Evolution and Current Status*, 107th Cong., 1st sess., April 4, 2001.
21. Barbara Larkin to James Sensenbrenner, December 22, 1998, available at http:// www.nasawatch.com/station.news.1998.html (accessed August 13, 2006).
22. White House, "President Bush Announces New Vision for Space Exploration Program," available at http://www.whitehouse.gov/news/releases/2004/01/20040114-3. html.
23. Leonard David, "NASA's Lunar Vision: The Devil's in the Details," Space.com, October 5, 2005, available at http://www.space.com/businesstechnology/technology/051005_nasa_details.html, (accessed July 12, 2006).
24. For further information, see Congressional Budget Office, "Budgetary Analysis of NASA's New Vision for Space Exploration."
25. Alex Wayne, "NASA and Earmarks Galore," *CQ Weekly*, November 27, 2004, 2807.
26. Mike Griffin, at NASA news conference; and "Exploration Systems Architecture Study" (transcript), September 19, 2005, available at http://www.spaceref.com/ news/viewsr.html?pid=18122 (accessed July 12, 2006).
27. "CEV: The Last Battlestar?" *Space Daily*, May 10, 2005.
28. NASA Final Request for Proposal (RFP), NNJ05111915R, Crew Exploration Vehicle (CEV), phase 2, January 11, 2006, available at http://www.spaceref.com/news/ viewsr.html?pid=19565 (accessed July 12, 2006).
29. "CEV."
30. John Kelley, "Griffin Talks About NASA's Goals and Challenges," *Florida Today*, November 16, 2005.
31. Traci Watson, "NASA Administrator Says Space Shuttle Was a Mistake," *USA Today*, September 27, 2005.

32. James Glanz, "Has the Shuttle Become NASA's '76 Dart?" *New York Times*, January 27, 2004, 1.
33. "Columbia Accident Investigation Board Report Excerpts," Space.com, available at http://www.space.com/missionlaunches/caib_details_030826.html (accessed June 21, 2006).
34. Warren E. Leary, "NASA Chief Sees Mandate for Bush Space Program," *New York Times*, November 23, 2004.
35. Michael Cabbage, "NASA Will Phase Out 100 Shuttle Jobs at KSC," *Orlando Sentinel*, November 23, 2004.
36. Commerce, Science and Transportation Committee, Subcommittee on Science and Space, 109th Cong., 1st sess., May 18, 2005.
37. Frank Mooring Jr., "One More Time: Hubble Program Planning Final Shuttle Mission as Early as 2007 to Maintain the Observatory," *Aviation Week & Space Technology*, July 11, 2005, 32.

4. The Militarization of Space

1. Michael T. Klare, "Endless Military Superiority," *Nation*, June 15, 2002.
2. White House, "Fact Sheet on National Space Policy," September 19, 1996, 3–4, available at http://www.ostp.gov/NSTC/html/pdd8.html (accessed July 31, 2006); Joint Chiefs of Staff, Department of Defense, *Joint Doctrine for Space Operations*, Joint Publication 3-14 (Washington, D.C.: Department of Defense, 2002), GL-2, available at http://www.dtic.mil/doctrine/jpoperationsseriespubs.htm (accessed June 20, 2006).
3. Joan Johnson-Freese and Roger Handberg, *Space, the Dormant Frontier: Changing the Paradigm for the 21st Century* (Westport, Conn.: Praeger, 1997). The major reports included National Commission on Space, *Pioneering the Space Frontier* (New York: Bantam, 1986); Sally K. Ride, "Leadership and America's Future in Space: Report to the NASA Administrator," available at http://history.nasa.gov/riderep/cover.htm (also known as the Ride Report after its chair and author, Dr. Sally Ride, the first American woman in space); Department of Defense, *Space Architecture Study* (1988); *Report of the Advisory Committee on the Future of the U.S. Space Program* (1990) (also known as the Augustine Report after its aerospace executive chair, Norm Augustine); *Synthesis Group Report* (1991), generated by a committee headed by retired Air Force general and astronaut Tom Stafford; National Space Council, *The Future of U.S. Space Launch Capability* (1992); Vice-President's Space Policy Advisory Board, *A Post Cold War Assessment of U.S. Space Policy* (1992); and National Research Council, *From Earth to Orbit* (1992).
4. Burt Rutan's successful claiming of the X Prize in 2004 with his *SpaceShipOne* may slowly see that changing, as the private sector becomes involved in manned spaceflight.
5. Atlas and Titan are built by Lockheed Martin; Delta, by Boeing.

6. Justin Ray, "Military Weather Satellite Repaired for October Liftoff," *Spaceflight Now*, June 27, 2002, available at spaceflightnow.com/titan/g9/020627fixed.html (accessed June 21, 2006).
7. R. C. Parkinson, "The Hidden Costs of Reliability and Failure in Launch Systems," *Journal of Reducing Space Mission Costs* 1, no. 3 (1998): 263–273.
8. General Chuck Horner, conversation with author, September 1994.
9. Air Force, "Evolved Expendable Launch Vehicle," fact sheet, available at http://www.au.af.mil/factsheets/eelv.htm (accessed October 23, 2006).
10. Marcia S. Smith, *Space Launch Vehicles: Government Activities, Commercial Competition, and Satellite Exports*, issue brief IB93062, prepared by Congressional Research Service, updated February 3, 2003, 5.
11. For a complete program description, see Randy Kendall, "EELV: The Next Stage of Space Launch," *Crosslink*, available at http://www.aero.org/publications/crosslink/winter2004/07.html (accessed June 21, 2006).
12. The Delta 4 "heavy" rocket, intended to replace the Titan 4 and provide the United States with heavy lift capabilities, was tested in December 2004. It failed to put its dummy satellite into the proper orbit.
13. John Kelley, "Rocket Program Deep in Red," *Florida Today*, October 10, 2004.
14. Raymond Decker, "Defense Space Activities: Continuation of Evolved Expendable Launch Vehicle Program's Progress to Date Subject to Some Uncertainty" (letter and presentation from General Accounting Office to Committee on Armed Services, Subcommittee on Strategic Forces, GAO-04-778R, June 24, 2004).
15. Ironically, in December 2004, the F-22 became a budget target to offset mounting deficits and the growing costs of the war in Iraq.
16. Jeffrey Lewis, "Lift-Off for Space Weapons? Implications of the Department of Defense's 2004 Budget Request for Space Weapons," *CISS*, July 21, 2003, available at www.cissm.umd.edu/papers/files/spaceweapons.pdf (accessed July 31, 2006).
17. Jefferson Morris, "Analysts: Lawmakers' Interest in Space Weapons Issue Heats Up," *Aerospace Daily and Defense Report*, April 7, 2005, 1.
18. Michael Bruno, "Defense Officials Outline Long-Range, Global-Strike Plans," *Aerospace Daily and Defense Report*, April 20, 2006, 1.
19. Adolfo J. Fernandez, *Military Role in Space Control: A Primer*," report RL32602, prepared by Congressional Research Service, September 23, 2004, 11–13.
20. However, while satellites carry fuel for maneuvering, they are like cars that leave the showroom with one tank of gas. There is no refueling in space. Using fuel for maneuvering directly reduces a satellite's useful lifetime, and so is done only on highest authority. Misperceptions about maneuvering are largely because of action movies, in which satellites seemingly are moved at the drop of a hat to get a better look at something.
21. Randall Correll, *Responsive Space: Transforming U.S. Space Capabilities* (Washington, D.C.: George Marshall Institute, 2004); Simon P. Worden and Randall Correll, "Responsive Space and Strategic Horizons," *Defense Horizons* 40 (2004): 1–8.
22. Fernandez, *Military Role in Space Control.*
23. Kinetic energy weapons are sometimes also referred to by the abbreviation KKV (kinetic kill vehicle).

24. Colin Robinson, "Defining Transformation," Center for Defense Information, June 25, 2002.
25. Donald Rumsfeld, "Twenty-first Century Transformation of the U.S. Armed Forces," available at http://www.defenselink.mil/speeches/2002/s20020131-secdef.html (accessed July 31, 2006).
26. Cost estimates vary from $4.5 billion to $14 billion, with Navy representatives claiming about $8.1 billion per ship. See "Costing the CVN-21: A DID Primer," *Defense Industry Daily*, December 19, 2005, available at www.defenseindustrydaily.com/2005/12/costing-the-cvn21-a-did-primer/index.php (accessed July 21, 2006).
27. David Talbot, "How Technology Failed in Iraq," *Technology Review* 107, no. 9 (2004): 36–42.
28. Quoted in Talbot, "How Technology Failed in Iraq," 38.
29. Matthew French, "Generals Point Ways to Better Blue Force," *Federal Computer Week*, October 21, 2003.
30. Wang Hucheng, "The U.S. Military's 'Soft Ribs' and Strategic Weaknesses," Beijing Xinhua Hong Kong Service, July 5, 2005.
31. Quoted in Patrick J. Garrity, "Implications of the Persian Gulf War for Regional Powers," *Washington Quarterly* 16, no. 3 (1993): 150; House Committee on National Security, *Quadrennial Defense Review*, 105th Cong., 1st sess., April 16, 1997.
32. Quoted in Christopher Hellman, "Last of the Big Time Spenders: U.S. Military Budget Still the World's Largest, and Growing," Center for Defense Information, available at http://www.cdi.org/issues/wme/spendersfy04.html (accessed June 21, 2006).
33. Robert Scales, "Transformation," *Armed Forces Journal* 142, no. 8 (2005): 22.
34. Scales, "Transformation," 22.
35. Michael Pillsbury, *China's Military Strategy Toward the U.S.: A View from* Open *Sources*, report prepared for the U.S.–China Economic and Security Review Commission, November 2001, 4, cited in U.S.–China Economic and Security Review Commission, "The National Security Implications of the Economic Relationship Between the United States and China," July 2002, 26n.47, available at http://www.uscc.gov/researchpapers/2000_2003/reports/ch1_02.htm.
36. Alastair Iain Johnstone, "Toward Contextualizing the Concept of Shashoujian," working paper, August 2002, available at www.people.fas.harvard.edu/~johnston/shashoujian.pdf (accessed July 31, 2006).
37. Scott Elliott, "Teets: America Must Reach for Space Dominance," *Air Force Link*, September 15, 2004, available at http://www.af.mil/news/story.asp?storyID=123008652 (accessed July 31, 2006).
38. Theresa Hitchens, *Future Security in Space: Charting a Cooperative Course* (Washington, D.C.: Center for Defense Information, 2004), 17.
39. Rumsfeld, "Twenty-first Century Transformation."
40. Quoted in Senate Armed Services Committee, Strategic Forces Subcommittee, *Hearings on FY 06 Defense Authorization Budget Request for Space Activities*, 109th Cong., 1st sess., March 16, 2005.
41. Department of Defense, "National Defense Strategy of the United States of America," March 2005, 12, available at http://www.globalsecurity.org/military/library/policy/dod/nds-usa_mar2005.htm (accessed June 21, 2006).

42. Thomas M. Nichols, "Anarchy and Order in the New Age of Prevention," *World Policy Journal* 22, no. 3 (2005): 2–23.
43. Some authors have also used the word "hedging" in the context of force planning to mean fully preparing—or perhaps overpreparing—for any possible tasking of military force. That definition, however, is also recognized to (deliberately) exaggerate others' capabilities and to be very costly. See Henry C. Bartlett, G. Paul Holman Jr., and Timothy E. Somes, *Strategy and Force Planning,* 4th ed. (Newport, R.I.: Naval War College Press, 2004), 29–34.
44. Department of Defense, *Strategic Deterrence Joint Operating Concept* (Offut Air Force Base, Neb.: United States Strategic Command, 2004), available at http://www.dtic.mil/futurejointwarfare/concepts/sd_joc_v1.doc (accessed June 21, 2006).
45. William D. Hartung and Frida Berrigan, "Lockheed Martin and the GOP: Profiteering and Pork Barrel Politics with a Purpose," Arms Trade Resource Center issue brief, World Policy Institute, July 31, 2000.
46. Lott also supported the $70 billion F-22 fighter when it was repeatedly challenged for being over budget and behind schedule, and the C-130 transport plane in quantities higher than requested by the Pentagon. Both are Lockheed programs.
47. Center for Public Integrity, "U.S. Contractors Reap the Windfalls of Post War Reconstruction," October 30, 2003, available at http://www.publicintegrity.org/wow/report.aspx?aid=65 (accessed July 31, 2006).
48. Hartung and Berrigan, "Lockheed Martin and the GOP."
49. Alfred Kaufman, "Caught in the Network," *Armed Forces Journal* 142, no. 7 (2005): 20.
50. Kaufman, "Caught in the Network," 21.
51. Charles Moskos, review of *The Culture of Defense*, by Christopher Van der Allen, *Parameters* 31, no. 3 (2001): 164–165.
52. Kaufman, "Caught in the Network," 21–22.
53. Christopher D. Van der Allen, *The Culture of Defense* (Lanham, Md.: Lexington Books, 2001), 113.

5. The Weaponization of Space

1. John T. Correll, "The Command of Space," *Air Force Magazine* 79, no. 10 (1996), available at www.afa.org/magazine/oct1996/1096edit.asp.
2. Christopher Booker, "Global Repositioning: Galileo Threatens 'Special Relationship' with U.S.," *Sunday Telegraph* (London), July 4, 2004.
3. Christopher Booker, "Star Wars: Continents Clash in Outer Space," *Sunday Telegraph*, October 31, 2004; Allister Heath, "U.S. Threatens to Take Space War to Third Dimension," *Business*, October 31, 2004; "U.S. Could Shoot Down EU Satellites if Used by Foes in Wartime," Agence France Presse, October 24, 2004.
4. Peter Brookes, "Militarizing Space," *New York Post*, June 7, 2005.
5. Joan Johnson-Freese, "The Viability of U.S. Anti-Satellite (ASTA) Policy: Moving Toward Space Control," occasional paper, no. 30, Institute for Security Studies, European Union, January 2000, 10–17.

6. Mark G. Markoff, "Disarmament and 'Peaceful Purposes' Provisions in the 1967 Outer Space Treaty," *Journal of Space Law* 4, no. 1 (1976): 11.
7. U.S. Space Command, *Vision for 2020* (Peterson Air Force Base, Colo.: U.S. Space Command, 1997), 3.
8. Quoted in Senate Committee on the Armed Services, Strategic Subcommittee, *Department of Defense Authorization for Appropriations for Fiscal Year 2000 and the Future Years Defense Program*, part 7, *Strategic*, 106th Cong., 1st sess., March 22, 1999, 329.
9. Wade Huntley, "Space Weaponization and Its Implications," in *The Future of Canadian-American Strategic Cooperation* (Ottawa: Royal Canadian Military Institute, 2005).
10. Five crisis scenarios and the impact of involving space weapons are examined in Jeffrey Lewis, *What If Space Were Weaponized: Possible Consequences for Crisis Scenarios* (Washington, D.C.: Center for Defense Information, 2004).
11. William B. Scott, "Wargames Zero In on Knotty Milspace Issues," *Aviation Week & Space Technology*, January 29, 2001, 52.
12. They were issued in February 1996 (late), May 1997, October 1998, December 1999, and December 2000.
13. "President Bush Delivers Graduation Speech at West Point," June 1, 2002, available at http://www.whitehouse.gov/news/releases/2002/06/20020601–3.html (accessed July 31, 2006).
14. John. L. Gaddis, "A Grand Strategy of Transformation," *Foreign Policy* 133 (2002): 50–58, available at http://www.ndu.edu/nduspeeches/gaddis.htm (accessed June 22, 2006).
15. Andrew J. Bacevich, "Bush's Grand Strategy," *American Conservative*, November 4, 2002, available at http://www.amconmag.com/2002/2002_11_04/bushs_grand_strategy.html (accessed July 31, 2006).
16. Joint Chiefs of Staff, *The National Military Strategy of the United States of America: A Strategy for Today; a Vision for Tomorrow* (Washington, D.C.: Joint Chiefs of Staff, 2005), available at www.defenselink.mil/news/Mar2005/d20050318nms.pdf (accessed June 22, 2006).
17. The potential deployment of TMD to Taiwan is a major political issue for East Asia, and an exception of the "general acceptance" premise.
18. Daniel Kleppner, Frederick K. Lamb, and David E. Mosher, "Boost Phase Defense Against Intercontinental Ballistic Missiles," *Physics Today* 57, no. 1 (2004): 30.
19. Major John E. Parkerson Jr., "International Legal Implications of the Strategic Defense Initiative," *Military Law Review* 116, no. 67 (1987): 99–100.
20. David Mosher, "Understanding the Extraordinary Cost of Missile Defense," *Arms Control Today* 30, no. 10 (2000): 9–16, available at http://www.rand.org/natsec_area/products/missiledefense.html (accessed June 22, 2006).
21. See, for example, Robert Shuey, *Missile Survey: Ballistic and Cruise Missiles of Foreign Countries*, report prepared by Congressional Research Service, February 10, 2000.
22. The issue of threat assessment credibility is addressed in "Is the USA's Threat Assessment Credible?" *Jane's Defense Weekly*, March 15, 2000.

23. The bipartisan Commission to Assess the Ballistic Missile Threat to the United States, which met in 1998 under the chairmanship of Donald Rumsfeld, has served as the primary impetus for congressional backing of NMD. It concluded that North Korea or Iran could develop an ICBM within five years with little warning. But it also noted that it would be easier for a state to develop a shorter-range, ship-launched ballistic missile than an ICBM. Also often cited is the 1999 national intelligence estimate: National Intelligence Council, "Foreign Missile Developments and the Ballistic Missile Threat to the United States Through 2015," available at http://www.fas.org/irp/threat/missile/nie99msl.htm (accessed July 31, 2006).
24. Among those is Joseph Cirincione, director of the Non-Proliferation Project, Carnegie Endowment for International Peace, who testified before Senate Committee on Governmental Affairs, Subcommittee on International Security, Proliferation and Federal Services, *Assessing the Ballistic Missile Threat*, 104th Cong., 2nd sess., February 9, 2000.
25. Donald Rumsfeld, press conference, Brussels, Belgium, June 6, 2002.
26. Walter Pincus and Dana Priest, "Some Iraq Analysts Felt Pressure from Cheney Visits," *Washington Post*, June 5, 2003.
27. From the summary given by NIC chair Richard Cooper, testimony before House Committee on National Security, *Emerging Missile Threats to North America During the Next 15 Years*, 104th Cong., 2nd sess., February 28, 1996.
28. Michael Dobbs, "How Politics Helped Redefine Threat," *Washington Post*, January 14, 2002.
29. North Korea claimed that it was a rocket launching a satellite. The United States and Japan considered it to be a missile. It is also now widely believed that the launch of Taepo Dong-1 in 1998 was an unsuccessful attempt to launch a satellite, as the North Koreans claimed.
30. Skeptics argue that North Korea has a long way to go before having ICBM capability. It has to develop a working missile, capable of carrying a militarily significant payload; develop a warhead that does not burn up during reentry; and be able to successfully mate the missile and the warhead—none of which are easy tasks.
31. For a more complete review of these events, see Joseph Cirincione, "Assessing the Assessment: The 1999 National Intelligence Estimate of the Ballistic Missile Threat," *Nonproliferation Review* 7, no. 1 (2000), available at http://www.ciaonet.org/olj/npr/npr_oocijo1.html (accessed July 31, 2006).
32. Bradley Graham, *Hit to Kill: The New Battle over Shielding America from Missile Attack* (New York: Public Affairs, 2001).
33. "Missile Wars: The Threat, Is It Real?" writer and prod. Sherry Jones, aired on *Frontline*, October 10, 2002.
34. Diane Vaughn, *The Challenger Launch Decision* (Chicago: University of Chicago Press, 1996). Vaughn outlines another case of changing organizational decision-making parameters. In that case, the parameters concerned what conditions were considered safe for a shuttle launch.
35. Quoted in Dobbs, "How Politics Helped Redefine Threat."
36. National Intelligence Council, "Foreign Missile Developments and the Ballistic Missile Threat to the United States Through 2015."

37. "Report of the Panel on Reducing Risk in Ballistic Missile Defense Flight Test Programs," February 27, 1998, available at http://www.fas.org/spp/starwars/program/welch/welch-1.htm (accessed July 21, 2006).
38. Union of Concerned Scientists, "Chronology of Missile Defense Tests," available at http://www.ucsusa.org/global_security/missile_defense/chronology-of-missile-defense-tests.html (accessed July 31, 2006).
39. William J. Broad, "M.I.T. Studies Accusations of Lies and Cover-up of Serious Flaws in Antimissile System," *New York Times*, January 2, 2003.
40. Government Accounting Office, "Missile Defense: Review of Allegations About an Early Missile Defense Flight Test," GAO-02-125 (letter and report to Senator Charles E. Grassley and Representative Howard L. Berman, February 2002), and *Missile Defense: Review of Results and Limitations of an Early Missile Defense Flight Test*, GAO-02-124 (Washington, D.C.: Government Accounting Office, 2002).
41. Government Accounting Office, "Missile Defense," 6, 13.
42. Council for a Livable World, "Pro-National Missile Defense Supporters' Shameless Distortion of Public Opinion," September 3, 1998, available at http://www.fas.org/news/usa/1998/09/980903-poll.htm (accessed July 31, 2006).
43. Pew Research Center, "Public Behind Bush on Key Foreign Issues; Modest Support for Missile Defense, No Panic," June 11, 2001, available at people-press.org (accessed July 31, 2006).
44. Matt Kelley, "Missile Intercept Test Will Succeed Regardless of Outcome, Pentagon Says," Associated Press, December 1, 2001.
45. "General: 'Glitch' Caused Missile Defense Test Failure," Cable News Network, January 13, 2005.
46. David Stout and John H. Cushman Jr., "Defense Missile for U.S. System Fails to Launch," *New York Times*, December 16, 2004.
47. Republican National Committee, "Protecting the Fellowship of Freedom from Weapons of Mass Destruction," available at http://www.cnn.com/ELECTION/2000/conventions/republican/features/platform.00/#47 (accessed July 31, 2006).
48. Joan Johnson-Freese and Jim Knox, "Preventing Armageddon: Hype, Heroes, Hollywood, and Reality," *IBIS* 53, no. 173 (2000): 173–179.
49. Government Accounting Office, "Status of Ballistic Missile Program in 2004," GAO-05-243, March 31, 2005; "Bush's Star War's Fantasy" [editorial], *New York Times*, June 25, 2005.
50. Bradley Graham, "Panel Faults Tactics in Rush to Install Antimissile System," *Washington Post*, June 10, 2005.
51. Cover, *Economist*, June 17–23, 2000.
52. The nickname appears in a forthcoming work by Tom Nichols, my colleague at the Naval War College.
53. David E. Sanger, "Bush Says U.S. May Have Been Able to Intercept North Korean Missile," *New York Times*, July 7, 2006, 5.
54. John Yaukey, "Missile Defense Gobbles Funds with Few Results," Gannett News Service, July 17, 2006.

55. "Hunter to Seek Increased BMD Funds After North Korea Missile Launches," *Space & Missile Defense Report*, July 17, 2006; Yaukey, "Missile Defense Gobbles Funds with Few Results."
56. The genesis of the MDA reaches back to the Reagan era. During that administration, the Strategic Defense Initiative Organization (SDIO) held the reins, but was then restructured and downsized as the Ballistic Missile Defense Organization (BMDO). The MDA took over from the BMDO.
57. Frances FitzGerald, "Indefensible," *New Yorker*, October 4, 2004, 33.
58. Dan L. Crippen, "Estimated Costs and Technical Characteristics of Selected Missile Defense Systems" (letter and cost estimates from Congressional Budget Office to Majority Leader Thomas A. Daschle, January 31, 2002).
59. John Isaacs, "Current Status of Missle Defense Program, Center for Arms Control and Non-Proliferation," May 2004, available at http://64.177.207.201/pages/16_571.html.
60. Emmanuel Evita, "U.S. Allies Look at Missile Defense," United Press International, December 18, 2004.
61. Douglas Fisher, "Canada Faces Reality Check," *London (Ontario) Free Press*, December 5, 2004.
62. "Canada Advised to Stay Out of U.S. Missile Defence Program," *Canadian Press*, February 24, 2006.
63. Thomas P. M. Barnett, "Mr. President, Here's How to Make Sense of Your Second Term, Secure Your Legacy, and Oh Yeah, Create a Future Worth Living," *Esquire*, February 2005, 90–93, 128, 130.
64. Robert Jervis, "Cooperation Under the Security Dilemma," *World Politics* 30, no. 2 (1978): 175.
65. Keir A. Lieber and Daryl G. Press, "The Rise of U.S. Nuclear Primacy," *Foreign Affairs* 85, no. 2 (2006): 42–54.
66. Peter Hayes, ed., *Space Power Interests* (Boulder, Colo.: Westview Press, 1996).
67. David Hardesty, "Space-Based Weapons: Long-Term Strategic Implications and Alternatives," *Naval War College Review* 58, no. 2 (2005): 45–68.
68. Quoted in J. Heronema, "AF Space Chief Calls War in Space Inevitable," *Space News*, August 12–18, 1996, 4.
69. Donald Rumsfeld, chair, "Report of the Commission to Assess United States National Security Space Management and Organization," January 11, 2001, executive summary, 10, available at http://www.defenselink.mil/pubs/space20010111.html (accessed July 6, 2006).
70. Michael Krepon, *Space Security or Space Weapons? A Guide to the Issues* (Washington, D.C.: Henry L. Stimson Center, 2005), 15.
71. Jervis, "Cooperation Under the Security Dilemma," 196.
72. United Nations, General Assembly, "General Assembly Adopts 49 Disarmament, International Security Texts on Recommendation of Its First Committee," press release GA/9829, November 20, 2000, available at http://www.un.org/News/Press/docs/2000/20001120.ga9829.doc.html (accessed June 23, 2006).
73. United Nations, General Assembly, "Resolution Adopted by the General Assembly, 58/36. Prevention of an Arms Race in Outer Space," A/RES/58/36, January 8, 2004.

74. Everett Dolman, *Astropolitik: Classical Geopolitics in the Space Age* (London: Cass, 2001).
75. Peter Hays, "Weapons in Space" (remarks in panel discussion at Non-Proliferation Conference, Carnegie Endowment for International Peace, Washington, D.C., November 15, 2002). Hays is a lieutenant colonel in the Air Force and the executive editor of *Joint Force Quarterly*.
76. Alfred Kaufman, "Caught in the Network," *Armed Forces Journal*, February 2005.
77. Larry Niven and Jerry Pournelle, *The Mote in God's Eye* (New York: Simon and Schuster, 1974), and *Lucifer's Hammer* (New York: Harper & Row, 1977).
78. Federation of American Scientists, *Ensuring America's Space Security* (Washington, D.C.: Federation of American Scientists, 2004).
79. Fred Donovan, "FAS: Spending on Space Weapons Creates 'Unstoppable' Momentum," *Aerospace Daily & Defense Report*, October 8, 2004, 2.
80. Dwayne A. Day, "General Power vs. Chicken Little," *Space Review*, May 23, 2005, available at http://www.thespacereview.com/article/379/1 (accessed June 23, 2006).
81. See, for example, Michael Krepon and Christopher Clay, *Space Assurance or Space Dominance? The Case Against Weaponizing Space* (Washington, D.C.: Henry L. Stimson Center, 2003), 29–35; and Robert Preston, Dana J. Johnson, Sean J. A. Edwards, Michael D. Miller, and Calvin Shipbaugh, *Space Weapons, Earth Wars* (Santa Monica, Calif.: RAND Corporation, 2002), chap. 3.
82. A Canadian study differentiates between space-based strike weapons (SBSW), systems operating from earth orbit capable of damaging or destroying either terrestrial or terrestrially launched objects passing through space, and earth-to-space and space-to-space weapons, or ASATs. See Simon Collard-Wexler, Jessy Cowan-Sharp, Sarah Estabrooks, Thomas Graham Jr., Robert Lawson, and William Marshall, *Space Security 2004* (Toronto: Northview Press, 2004), 137.
83. For information on space weapons and the U.S. budget, see Jeffrey Lewis, "Space Weapons in the 2005 U.S. Defense Budget Request" (paper presented at the Outer Space and Global Security Conference, Geneva, Switzerland, March 25–26, 2004).
84. Theresa Hitchens, "Development in Military Space: Movement Toward Space Weapons?" (paper written with support from the Carnegie Corporation, Center for Defense Information, Washington, D.C, October 2003), 3–5; Jeffrey Lewis, "Lift Off for Space Weapons: Implications of the Department of Defense's 2004 Budget Request for Space Militarization" (paper, Center for International and Security Studies, University of Maryland, College Park, July 21, 2003), 15.
85. Office of the Press Secretary, White House, "National Policy on Ballistic Missile Defense Fact Sheet," press release, May 20, 2003.
86. Jeremy Singer, "Everett Defends Need to Defend U.S. Spacecraft," *Space News*, October 16, 2006, 10.
87. Lewis, "Lift Off for Space Weapons," 5.
88. Matt Bille, Robyn Kane, and Drew Cox, "Military Microsatellites: Matching Requirements and Technology" (paper presented at the AIAA Space 2000 Conference and Exhibition, Long Beach, Calif., September 19–21, 2000), cited in Hitchens, "Developments in Military Space," 9.

89. Quoted in Elaine M. Grossman and Keith J. Costa, "Small, Experimental Satellite May Offer More than Meets the Eye," *Inside the Pentagon*, December 4, 2003.
90. David Wright, Laura Grego, and Lisbeth Gronlund, *The Physics of Space Security: A Reference Manual* (Cambridge, Mass.: American Academy of Arts and Sciences, 2005).
91. Bruce M. DeBlois, Richard L. Garwin, R. Scott Kemp, and Jeremy C. Marwell, "Space Weapons: Crossing the U.S. Rubicon," *International Security* 29, no. 2 (2004): 58.

6. The Politicization of the U.S. Aerospace Industry

1. David H. Napier, *2004 Year-End Review and 2005 Forecast—An Analysis* (Arlington, Va.: Aerospace Industries Association, 2004), 1.
2. Futron Corporation, "2001–2002 Satellite Industry Indicators Survey," slide 5, available at www.futron.com/pdf/ Satellite%20Industry%20Indicators%20Survey-02.pdf (accessed July 31, 2006).
3. Cox was subsequently tapped in June 2005 by President George W. Bush to be chairman of the Securities and Exchange Commission.
4. Chalmers Johnson, "In Search of a New Cold War," *Bulletin of the Atomic Scientists* 55, no. 5 (1999): 44–51.
5. The information on launch failures, while embarrassing and sensitive to the manufacturers and agencies involved, is certainly not classified. A good description of all the world's launch vehicles and their detailed histories can be found in Steven J. Isakowitz, Joseph P. Hopkins Jr., and Joshua B. Hopkins, *International Reference Guide to Space Launch Systems*, 3rd ed. (Reston, Va.: American Institute of Aeronautics and Astronautics, 1999).
6. Intelsat has launched nine series of satellites, designated the Intelsat I through Intelsat IX. The tenth series is under construction at Astrium, in Toulouse, France. There have been up to fifteen satellites in each series. Individual satellites in each series are given a unique Arabic number; thus *Intelsat 603* refers to the third satellite in the Intelsat VI series.
7. This was by no means the first instance of a communications breakdown in the space program. See Phillip K. Thompkins, *Organizational Communication Imperatives: Lessons of the Space Program* (Los Angeles: Roxbury, 1993).
8. Brazing is the process of joining metals using heat and a filler metal with a melting temperature below the melting point of the materials being joined. In the case of the RL-10, the drawing allowed for small voids in the filler material, but was misinterpreted to allow one large void instead of several smaller voids.
9. Russian Launch Failure Review Board, "PROTON—Analysis of Causes of 5 July 1999 Proton/Raduga Launch Failure," released by the Khrunichev Space Center into the public domain on August 3, 1999.
10. "Report by Inter-agency Board for Investigating Causes of Failure of Proton LV + 11C861 Upper Stage Stack During Express A #1 Spacecraft Launch Mission on October, 27 1999," released by the Khrunichev Space Center into the public domain on January 7, 2000.

11. Quoted in Senate Committee on Banking, Subcommittee on International Trade and Finance, *Rethinking Export Controls*, 106th Cong., 1st sess., March 16, 1999.
12. William New, "Exports: Juster Notes Changes in Commerce, Security During Tenure," *National Journal's Technology Daily*, P.M. edition, December 21, 2004.
13. House Select Committee on U.S. National Security and Military/Commercial Concerns with the People's Republic of China, *The U.S. National Security and Military/Commercial Concerns with the People's Republic of China*, 105th Cong., 2nd sess., January 3, 1999, chap. 5, 4.
14. Joan Johnson-Freese, "Becoming Chinese: Or, How U.S. Satellite Export Policy Threatens National Security," *Space Times*, January–February 2001, and "Alice in Licenseland: U.S. Satellite Export Controls Since 1990," *Space Policy* 16, no. 3 (2000): 195–204.
15. Page numbers can differ, depending on which appendix information is included in the count. Usually the classified version is stated as 1,000 pages and the declassified version at about 700 pages, though 860 is also a common page number for the declassified version.
16. Andrea Mitchell, interview with Christopher Cox, *NBC Nightly News*, May 21, 1999.
17. The influence of the press on the events surrounding the Cox Committee investigation should not be underestimated. The *New York Times* staff, with a special mention for Jeff Gerth, was awarded a Pulitzer Prize for a series of articles that, according to the award citation, reported that despite national security risks, two aerospace companies had improperly shared expertise with China with the approval of the United States government. See Alex Kuczynski, "Teacher Turned Playwright Is Among the Winners of 22 Pulitzer Prizes," *New York Times*, April 13, 1999, B9.
18. "China Syndrome" [editorial], *Wall Street Journal*, April 14, 1998. President Clinton approved the launch of a satellite. The transfer of missile guidance technology was neither requested nor granted. The article's illogic (how can a technology be transferred a second time?) is perhaps encouraged by the language of the required TAA, which mirrors ITAR definitions of technology transfer to describe the routine sharing of information necessary for the business transaction in question.
19. John M. Broder, "China Issues Resists Usual White House Defenses," *New York Times*, May 20, 1998, A16.
20. Leon Hadar, "Gosh! The Chinese Have Spies?" *Business Times* (Singapore), May 27, 1999.
21. See, for example, "China Buys . . ." [editorial], *Wall Street Journal*, March 11, 1999, A22.
22. Quoted in Walter Pincus, "Hill Report on Chinese Spying Faulted," *Washington Post*, December 15, 1999, A16.
23. Tom Plate, "Cox Report Was 'An Exercise in Amateur-Hour Paranoia,'" *Los Angeles Times*, July 21, 1999, B7. See also Jonathan Pollack, "The Cox Report's 'Dirty Little Secret,'" *Arms Control Today* 29, no. 3 (1999).
24. Richard L. Garwin, "Why China Won't Build US Warheads," *Arms Control Today* 29, no. 3 (1999); Lars-Erik Nelson, "Washington: The Yellow Peril," *New York Review of Books*, July 15, 1999.

25. Senate Select Committee on Intelligence, *Report on Impacts on U.S. National Security of Advanced Satellite Technology Transfers to the People's Republic of China (PRC) and Report on the PRC's Efforts to Influence U.S. Policy*, 105th Cong., 2nd sess., May 5, 1999.
26. Central Intelligence Agency, "The Intelligence Community Damage Assessment on the Implications of China's Acquisition of US Nuclear Weapons Information on the Development of Future Chinese Weapons," Key Findings, April 21, 1999, available at http://www.fas.org/sgp/news/dci042199.html (accessed July 31, 2006).
27. Joseph Cirincione, "The Political and Strategic Imperatives of National Missile Defense" (paper presented at the seventh ISODARCO Beijing Seminar on Arms Control, Xi'an, China, October 8–12, 2000), available at http://www.ceip.org/files/publications/imperativesnmd.asp (accessed August 13, 2006).
28. Peter B. deSelding, "Export Issue Touches Europe," *Space News*, March 8, 1999, 20.
29. Mark Stokes, *China's Strategic Modernization* (Carlisle, Pa.: Strategic Studies Institute, 1999), 26.
30. Senate Committee on Banking, Housing and Urban Affairs, *Hearings on the Export Administration Act*, 106th Cong., 1st sess., June 24, 1999.
31. Bryan Bender, "US Industry to Propose Regime for Public–Private Export Control," *Jane's Defense Weekly*, May 12, 1999.
32. David S. Cloud and Helene Cooper, "US Says China-Satellite Rejection was One-time Event, but Chill May Result," *Wall Street Journal*, February 24, 1999, A4.
33. Leslie Chang, "US Firms Rue Negative Effects of Cox Report—Stricter Scrutiny of Ties to China May Threaten Lucrative Contracts," *Asian Wall Street Journal*, May 27, 1999, 1.
34. Susan V. Lawrence, "Clipping Their Wings," *Far Eastern Economic Review*, April 8, 1999, 20.
35. "Intelsat Might Move Out of US," *Space News*, October 19, 1999, 6. There is still an Intelsat office in Washington.
36. Peter B. deSelding, "US Export Rules Frustrate Germans," *Space News*, July 5, 1999, 1, 20.
37. Foreign companies negatively affected by the new licensing process through, for example, being unable to get requisite information for a successful launch may not feel so constrained about taking judicial action. According to Simon D. Clapham, head of the space department at Marham Space Consortium, London, "[I]f the State Department does not allow us to have that information, then I suppose a court of law will have to sort this out" (interview with Peter deSelding, *Space News*, November 22, 1999, 22).
38. *Explanatory Statement Related to HR 3427*, House Report 106–479, *Congressional Record*, 106th Cong., 1st sess., November 17, 1999, 145, H12586.
39. Joseph Anselmo, "Congress Seeks Fix to Export Quagmire," *Aviation Week & Space Technology*, May 10, 1999, 72.
40. Anselmo, "Congress Seeks Fix to Export Quagmire," 72.
41. Anselmo, "Congress Seeks Fix to Export Quagmire," 72.

42. Web site of Dana Rohrabacher, available at http://rohrabacher.house.gov/Committees/ (accessed July 24, 2006).
43. Quoted in Eric Schmitt, "Chinese Suddenly Improve Rocket Safety, Expert Says," *New York Times*, June 18, 1998, 6. The witness making the assertion to which Rohrabacher was responding was Joan Johnson-Freese.
44. Quoted in Senate Committee on Foreign Relations, *International Economic Policy, Export and Trade Promotion: US Satellite Export*, 106th Cong., 1st sess., June 24, 1999, 4.
45. Cynthia Cotts, "What? No Smoking Gun?" *Village Voice*, September 15–19, 1999.
46. Laurie Sullivan, "Electronics Execs Chafe Under Tough U.S. Export Rules," *EETimes*, September 25, 2003.
47. Available at www.cpb.gov (accessed July 31, 2006).
48. George Abbey and Neal Lane, *U.S. Space Policy: Challenges and Opportunities* (Cambridge, Mass.: American Academy of Arts and Sciences, 2005).
49. Dave Weldon, "Export Control Policy Hampering U.S. Competitiveness," *Space News*, June 27, 2005.
50. "Licensing Jurisdiction for 'Space Qualified' Items and Telecommunications Items for Use on Board Satellites: Final Rule, Amendments of the United States Munitions List," *Federal Register* 67, no. 184 (September 23, 2002), 59721–59733, available at http://frwebgate.access.gpo.gov/cgi-bin/getdoc.cgi?dbname=2002_register&docid=02-23713-filed (accessed July 31, 2006).

7. The Ambitions of Europe

1. Stefano Silvestri, "Space and Security Policy in Europe," executive summary, occasional paper, no. 48, Institute for Security Studies, European Union, Paris, December 2003, 10.
2. Much of this history is taken from Joan Johnson-Freese, *Changing Patterns of International Cooperation in Space* (Malabar, Fla.: Orbit Books, 1990), 11–13, 25–26.
3. Reimar Luest, "Cooperation Between Europe and the United States in Space" (paper presented at the Association of American Universities, the Fulbright Fortieth Anniversary Lecture, Washington, D.C., April 6, 1987), 20.
4. Arnold Frutkin, *International Cooperation in Space* (Englewood Cliffs, N.J.: Prentice-Hall, 1965), 133.
5. Arturo Russo, "Launching the European Telecommunications Satellite Program," in *Beyond the Ionosphere: The Development of Satellite Communications*, NASA History Series, SP-4217 (Washington, D.C.: NASA History Office, 1997), available at http://history.nasa.gov/SP-4217/ch10.htm (accessed July 28, 2006).
6. Johnson-Freese, *Changing Patterns of International Cooperation in Space*, 26–30.
7. Johnson-Freese, *Changing Patterns of International Cooperation in Space*, 83–91.

8. Hermann Strub, "Columbus—The European Contribution to the International Space Station," in *Proceedings of the International Symposium on Europe in Space—The Manned Space System*, ESA SP-277 (Paris: European Space Agency, 1988), 50–51.
9. Wulf von Kries, "Flunking on Space Station Cooperation?" *Space Policy* 3, no. 1 (1987): 11.
10. NASA Advisory Council, Task Force on International Relations in Space, *International Space Policy for the 1990s and Beyond* (Washington, D.C.: NASA, 1987), 30.
11. Chris Morris, "EU Rebuffs US over Satellite Project," BBC News, March 8, 2002, available at http://news.bbc.co.uk/1/hi/world/europe/1862779.stm (accessed June 23, 2006).
12. Daniel Keohane, introduction to Carl Bildt, Mike Dillon, Daniel Keohane, Xavier Pasco, and Tomas Valasek, *Europe in Space* (London: Centre for European Reform, 2004), 4–5.
13. Joan Johnson-Freese and Lance Gatling, "Security Implications of Japan's Information Gathering Satellite System," *Intelligence and National Security* 19, no. 3 (2004): 538–552.
14. Silvestri, "Space and Security Policy in Europe," 3.
15. Silvestri, "Space and Security Policy in Europe," 9.
16. This is known as the "Lisbon Process." See Carl Bildt and Mike Dillon, "Europe's Final Frontier," in Bildt et al., *Europe in Space*, 7.
17. European Commission, "Progress Report on the Galileo Research Programme," report to the Commission of the European Communities, Brussels, February 18, 2004, available at http://www.europa.eu.int/comm/dgs/energy_transport/galileo/doc/com_2004_0112_en.pdf (accessed June 23, 2006).
18. Marcia S. Smith, *U.S. Space Programs: Civilian, Military, and Commercial*, issue brief IB92011, prepared by Congressional Research Service, May 29, 2003.
19. Keohane, introduction, 3.
20. At the Barcelona summit in March 2001, the EU governments agreed that European R&D spending should increase from the current 2 percent of GDP to 3 percent by 2010. See Bildt and Dillon, "Europe's Final Frontier," 7.
21. European Commission, "Progress Report on the Galileo Research Programme," 5.
22. Silvestri, "Space and Security Policy in Europe," 20.
23. Xavier Pasco, "Ready for Take-Off? European Defence and Space Technology," in Bildt et al., *Europe in Space*, 28.
24. For a more extended look at relations between Europe and China, see Joan Johnson-Freese and Andrew Erickson, "The Emerging China–EU Space Partnership: A Geo-Technological Balancer," *Space Policy* 22, no. 1 (2006): 12–22.
25. Assembly of Western European Union, Technological and Aerospace Committee, "Aerospace Cooperation Between Europe and China," document A/1853, June 3, 2004, 2.
26. Web site of the European Space Agency, available at http://www.esa.int/esaSC/120381_index_0_m.html (accessed July 29, 2006).

27. Assembly of Western European Union, "Aerospace Cooperation Between Europe and China," 8.
28. Assembly of Western European Union, "Aerospace Cooperation Between Europe and China," 2.
29. Since NASA administrator Michael Griffin's trip to China in September 2006, the possibility of China's inclusion in the ISS has been quietly discussed in space circles. In an interesting turn of events, there is concern among the partners that the United States will use China's participation as an excuse to again rewrite timetables and abrogate its responsibilities to the ISS.
30. Assembly of Western European Union, "Aerospace Cooperation Between Europe and China," 13.

8. The Ambitions of China

1. State Council, Information Office, "China's Space Activities," November 22, 2000, available at http://www.spaceref.com/china/china.white.paper.nov.22.2000.html (accessed July 31, 2006).
2. State Council, Information Office, "China's Space Activities in 2006," October 12, 2006, available at www.cnsa.gov.cn/n6157091/n639462/79381.html.
3. Thomas P. M. Barnett, "The Pentagon's New Map," *Esquire*, March 2003, 175, and *The Pentagon's New Map* (New York: Putnam, 2004).
4. Henry Kissinger, "China: Containment Won't Work," *Washington Post*, June 13, 2005, 19; Benjamin Schwartz, "Managing China's Rise," *Atlantic Monthly*, June 2005; David Shambaugh, "The New Strategic Triangle: U.S. and European Reactions to China's Rise," *Washington Quarterly* 28, no. 3 (2005): 9–25.
5. Pew Global Attitudes Project, *U.S. Image Up Slightly, But Still Negative: American Character Gets Mixed Reviews* (Washington, D.C.: Pew Global Attitudes Project, 2005), 2, cited in Andrew Kohurt and Carolyn Funk, "Poll: In Wake of Iraq War, Allies Prefer China to U.S.," June 24, 2005, available at www.pewglobal.org/reports/pdf/247.pdf (accessed August 13, 2006).
6. The rationale for large-scale visa rejection is that China's "national-level intelligence services employs a full range of collection methodologies, from targeting of well-placed foreign government officials, senior scientists and businessmen to the exploitation of academic activities, student populations and private businesses" (Michelle Van Cleave, quoted in Bill Gertz, "China a 'Central' Spying Threat," *Washington Times*, September 29, 2005, 4).
7. Cassie Biggs, "U.S. Lawmaker: Bush Wants to Discuss Space Cooperation with China," Associated Press, January 11, 2006.
8. Christopher Hill, assistant secretary of state for East Asian and Pacific affairs, testimony before Senate Foreign Relations Committee, Subcommittee on East Asian and Pacific Affairs, *Emergence of China in the Asia-Pacific, Economic and Security Consequences for the United States*, 109th Cong., 1st sess., July 7, 2005.
9. See, for example, Kissinger, "China."

10. State Council, "China's Space Activities."
11. Joan Johnson-Freese, "Space Wei Qi: The Launch of *Shenzhou V*," *Naval War College Review* 57, no. 2 (2004): 121–145, and "China's Manned Space Program: Sun Tzu or Apollo Redux?" *Naval War College Review* 56, no. 3 (2003): 51–71.
12. The Chinese number their programs. The first two numbers indicate that it was started in 1992.
13. Antoaneta Bezlova, "Science: By Launching a 'Taikonaut,' China Enters the Space Race," *Global Information Network*, October 15, 2003, 1.
14. Yang also carried other commemorative items, including crop seeds from Taiwan.
15. "India Can Match China's Space Programme," *Times of India*, October 16, 2003.
16. "Indian Prime Minister Hails Chinese Manned Space Flight," Agence France Press, October 18, 2003.
17. "China's Launch of Manned Spacecraft Welcomed in Japan," Japan Economic Newswire, October 15, 2003.
18. "China's Launch of Manned Spacecraft Welcomed in Japan."
19. "Rocket Carrying Two Spy Satellite Destroyed," *Mainichi News*, November 29, 2003.
20. Kyoko Takita, "Japan Must Look Again at Space-Market Strategy," *Yomiuri Shimbun*, November 14, 2003, and "Lack of Expertise Dogs Space Program," *Yomiuri Shimbun*, November 24, 2003; Yohio Shioya, "A Political Step to the Stars," *Nihon Keizai Shimbun*, November 3, 2003.
21. Luo Ge, "The Global Space Agenda: China" (paper presented at the Center for Strategic and International Studies, Human Space Exploration Initiative, Washington, D.C., April 3, 2006).
22. "Chinese Annual Space Budget Exceeds Two Billion Dollars," Spacedaily.com, October 12, 2006, available at http://www.spacedaily.com/reports/Chinese_Annual_Space_Budget_Exceeds_Two_Billion_Dollars_999.html.
23. State Council, "China's Space Activities."
24. State Council, "China's Space Activities in 2006."
25. Particularly during the Apollo program, NASA was known for its youthful culture. Since the 1990s, however, NASA has faced difficult workforce issues. See Sean O'Keefe, "NASA Workforce Issues," statement to Senate Committee on Governmental Affairs, 108th Cong., 1st sess., March 6, 2003.
26. U.S. Congress, Office of Technology Assessment, *Reducing Launch Operations Costs: New Technologies and Practices*, OTA-TM-ISC-28 (Washington, D.C.: Government Printing Office, 1988), 39–40.
27. Quoted in John Pomfret, "Chinese Officials Plot a Path in Space," *Washington Post*, October 16, 2003.
28. See the description in Ben Iannotta, "China's Divine Craft," *Aerospace America*, April 2001.
29. CAC was also sometimes known as CASC. Chinese bureaucracies and organizations are regularly reorganized and renamed; though personnel and buildings rarely change, some reorganizations are more substantive than others. The more recent efforts are seen as real efforts toward more effective management,

while avoiding the kind of economic collapse the Chinese viewed with horror in the early 1990s when "reform" occurred in the former Soviet Union. Subsequently, however, acronyms can become particularly confusing. CASTC is often still referred to as CASC, referencing its predecessor organization CAC/CASC.

30. CASIC initially was called China Aerospace Machinery and Electronics Corporation (CAMEC).
31. Sichuan Space Industry Corporation and Xian Space Science & Technology Industry Corporation.
32. Federation of American Scientists, "Chinese Aerospace Corporation—CASC—China Nuclear Forces," available at http://www.fas.org/nuke/guide/china/contractor/casc.htm (accessed June 23, 2006).
33. "Chinese Aerospace Corporation Meeting Discusses Development," Xinhua News Agency, reported in BBC Worldwide Monitoring, August 13, 2002.
34. Bifurcation was not an issue for Japan and some European countries because they specifically chose not to pursue military space endeavors. For others—France, for example—the military has and continues to play a significant role in its space program. The French Ministry of Defense has oversight responsibility for Centre National d'Études Spatial (CNES), the French space agency, and directly contributes to its budget for programs such as the Helios military observation satellites.
35. Hence, Chinese launchers are referenced as Long March (LM) or Chang Zheng (CZ), and strategic missiles as Dong Feng (DF). Many Chinese space systems (and institutions) have multiple and sometimes "classified" designations, for internal and external use. Sometimes these multiple designations are deliberate and sometimes merely a function of history. Individuals within portions of CASC, for example, often still refer to numerical academy and institution functions that have long since changed, many times. This can be a source of constant confusion for outsiders.
36. Craig Covault, "Shenzhou Solos," *Aviation Week & Space Technology*, October 20, 2003, 22.
37. Wu Tse and Li Tsinlung were sent to the cosmonaut training center in Zvezdny Gorodok, Russia.
38. Richard D. Fisher, testimony before House Armed Services Committee, 109th Cong., 1st sess., July 27, 2005.
39. "Space Mission Generating National Pride but Also Criticism," AsiaNews.it, October 13, 2005.
40. China is credited with being far more astute at utilizing open-source material than the United States, to the extent that its abilities are considered a serious security threat and, though not illegal, part of Chinese espionage efforts. See "U.S. Grapples with Intense Chinese Spying," United Press International, June 27, 2005.
41. Larry M. Wortzel, "China and the Battlefield in Space," WebMemo no. 346, October 15, 2003.
42. Fisher, testimony before House Armed Services Committee.
43. Gertz, "China a 'Central' Spying Threat," 4.

44. Defense Science Board, "Report of the Defense Science Board Task Force on Strategic Communication," 2004, 173, available at www.acq.osd.mil/dsb/reports/2004-09-Strategic_Communication.pdf.
45. J. Michael Waller, "PLA Revises the Art of War," *Insight on the News*, February 28, 2000.
46. Gregory Kulacki and David Wright, "New Questions About U.S. Intelligence on China," September 15, 2005, available at http://www.ucsusa.org/global_security/china/new-questions-about-us-intelligence-on-china.html (accessed June 23, 2006).
47. For technical parameters, see Hu Xiao-gong, Huang Cheng, and Huang Yong, "Simulation of Precise Orbit Determination for a Lunar Orbiter," *Chinese Astronomy and Astrophysics* 46, no. 2 (2005): 186–195.
48. Quoted in "China Eyes 2017 Moon Landing," Reuters, November 4, 2005.
49. Charles Vick, "China and Russia Challenging the Space Leadership of the United States," GlobalSecurity.org, October 14, 2005, available at http://www.globalsecurity.org/space/library/report/2005/051014-prc-cpv.htm (accessed June 23, 2006).
50. Jacqueline Newmyer, testimony before U.S.–China Economic and Security Review Commission, *Filling a Gap in Our Understanding of Chinese Strategy*, 109th Cong., 2nd sess., March 16, 2006.
51. For comparison, the U.S. Atlas V, part of the EELV family, can lift up to 29,500 pounds, or more than 14 tons.
52. Nan Li, "The PLA's Evolving Campaign Doctrine and Strategy," in James C. Mulvenon and Richard H. Yang, eds., *The People's Liberation Army in the Information Age* (Santa Monica, Calif.: RAND Corporation, 1999), 146–174.
53. Liu Shiquan, "A New Type of 'Soft Kill' Weapon: The Electromagnetic Pulse Warhead," *Hubei hangtian jishu* [Hubei Space Technology], May 1997, 46–48.
54. Donald Rumsfeld, chair, "Report of the Commission to Assess United States National Security Space Management and Organization," available at http://www.defenselink.mil/pubs/space20010111.html (accessed July 6, 2006).
55. See, for example, Thomas E. Ricks, "Space Is Playing Field for Newest War Game," *Washington Post*, January 29, 2001, 1.
56. "Advantages of 'Shenzhou' Spacecraft, 'Long-March' Carrier Rocket," People's Daily Online, English version available at http://fpeng.peopledaily.com.cn/home.shtml.
57. James Oberg, "China's Great Leap Upward," ScientificAmerican.com, September 15, 2003, available at http://www.sciam.com (accessed June 23, 2006), which refers to the June 2000 issue of *Xiandai Bingqi*, the monthly journal of a military technology research institute.
58. Swedish engineer and analyst Sven Grahn provides interesting tracking data on the *Shenzhou V* mission in "The Flight of Shenzhou 5," available at http://www.svengrahn.pp.se/histind/China12/Shenzhou5.html (accessed June 23, 2006). Also, in November 2003, the republic of Kiribati, the location of one of China's two external tracking sites, diplomatically recognized Taiwan. That created significant issues for China, because if it subsequently broke relations with Kiribati in response, it risked forfeiting the tracking site. See Philip P. Pan, "Tiny Repub-

lic Embraces Taiwan, and China Feels Betrayed," *Washington Post*, November 27, 2003, A15.

59. The Air Force's Manned Orbiting Laboratory, for example, was canceled in the 1960s.
60. Craig Covault, "Chinese Milspace Ops," *Aviation Week & Space Technology*, October 20, 2003, 26.
61. Quoted in Marc Selinger, "Rep. Rohrabacher Sees Progress in Bid to Boost Foreign Role in ISS," *Aerospace Daily*, August 30, 2001, 3.
62. Franklin Kramer, testimony before House Armed Services Committee, 109th Cong., 1st sess., July 27, 2005.
63. Fisher, testimony before House Armed Services Committee.
64. Robert Preston and John Baker, "Space Challenges," in *Strategic Appraisal: United States Air and Space Power in the 21st Century* (Santa Monica, Calif.: RAND Corporation, 2002), 145.
65. Robert Jervis, "Cooperation Under the Security Dilemma," *World Politics* 30, no. 2 (1978): 211.

9. Avoiding a Clash of Ambitions

1. Joan Johnson-Freese, "The New U.S. Space Policy: A Turn Toward Militancy?" *Issues in Science and Technology* 23, no. 2 (2007): 33–36.
2. National Space Policy, PDD/NSTC-8, September 1996, available at www.fas.org/spp/military/docops/national/nstc-8.htm (accessed July 31, 2006).
3. George Abbey and Neal Lane, *United States Space Policy: Challenges and Opportunities* (Cambridge, Mass.: Academy of Arts and Sciences, 2005).
4. United States Mission to the United Nations, press release, no. 285(06)1, October 11, 2006.
5. Jerome Bernard, "New US Space Policy Targets Rivals' Capabilities," Agence France-Press, October 19, 2006.
6. Simon Collard-Wexler, Jessy Cowan-Sharp, Sarah Estabrooks, Thomas Graham Jr., Robert Lawson, and William Marshall, *Space Security 2004* (Toronto: Northview Press, 2004), v.
7. Collard-Wexler et al., *Space Security 2004*, v.
8. Thomas M. Nichols, *Winning the World: Lessons for America's Future from the Cold War* (Westport, Conn.: Praeger, 2002), 165–204.
9. Pew Global Attitudes Project, *U.S. Image Up Slightly, But Still Negative: American Character Gets Mixed Reviews* (Washington, D.C.: Pew Global Attitudes Project, 2005), 23.
10. Dwayne A. Day, "General Power vs. Chicken Little," *Space Review*, May 23, 2005, available at http://www.thespacereview.com/article/379/1 (accessed June 23, 2006).
11. Day, "General Power vs. Chicken Little," 11.
12. Joan Johnson-Freese, "Strategic Communication with China: What Message About Space?" *China Security* 2 (2006): 37–57.
13. Jeremy Singer, "Cartwright Seeks Closer Ties with China, Russia: Cooperation,

Personnel Exchanges Seen as Helpful in Avoiding Misunderstandings," *Space News*, October 16, 2006, 14.

14. Pew Global Attitudes Project, "Bush Unpopular in Europe, Seen as Uniliateralist," August 15, 2001, available at http://pewglobal.org/reports/display.php?PageID = 801 (accessed June 23, 2006).
15. Robert Jervis, "Cooperation Under the Security Dilemma," *World Politics* 30, no. 2 (1978): 187.
16. James J. Wirtz and James A Russell, "A Quiet Revolution: Nuclear Strategy for the 21st Century," *Joint Forces Quarterly* 33 (2002–2003): 11.
17. Jervis, "Cooperation Under the Security Dilemma," 181.
18. Theresa Hitchens, *Future Security in Space: Charting a Cooperative Course* (Washington, D.C.: Center for Defense Information, 2004), 53–55.
19. The work of the Reconsidering the Rules for Space Project is available at http://www.cissm.umd.edu/projects/space.php (accessed July 31, 2006).
20. "A Model Code of Conduct for the Prevention of Incidents and Dangerous Military Practices in Outer Space," available at www.stimson.org/pub.cfm?id = 106 (accessed July 31, 2006); Michael Krepon and Michael Heller, "A Model Code of Conduct for Space Assurance," *Disarmament Diplomacy* 77 (June 2004), available at http://www.acronym.org.uk/dd/dd77/77mkmh.htm (accessed July 31, 2006).
21. Singer, "Cartwright Seeks Closer Ties with China, Russia."
22. Traci Watson and Dan Vergano, "NASA Chief Says He Can't Promise Space Station's Completion," *USA Today*, June 21, 2005.
23. White House, "Press Briefing by Deputy National Security Advisor Faryar Shirzad and National Security Council Acting Senior Director Dennis Wilder on the President's Meetings with President Hu of the People's Republic of China," April 20, 2006, available at http://www.whitehouse.gov/news/releases/2006/04/20060420-7.html (accessed August 15, 2006).
24. American Institute of Aeronautics and Astronautics, *From Challenges to Solutions* (report of the seventh International Space Cooperation Workshop, Anchorage, Alaska, May 2–6, 2004). See also Peggy Finarelli and Ian Pryke, "Optimizing Space Exploration Through International Cooperation" (paper presented following the seventh International Space Cooperation Workshop, Anchorage, Alaska, May 2–6, 2004).
25. Sarah Chankin-Gould and Ivan Oelrich, "Double-Edge Shield: The United States Is Considering Relaxing Export Controls So It Can Share Missile Defense Technology with Its Friends—A Move that Could Help Its Enemies," *Bulletin of the Atomic Scientists*, May 1, 2005, 36.
26. "Steidle Named AIA Vice President of International Affairs," *PR Newswire US*, June 16, 2005.
27. Jon Howland, "Foes See U.S. Satellite Dependence as Vulnerable Asymmetric Target," JINSA Online, December 4, 2003, 14.
28. Office of the Secretary of Defense, "The Military Power of the People's Republic of China," July 20, 2005, available at http://www.defenselink.mil/news/Jun2000/china06222000.htm.
29. Azar Nafisi, *Reading Lolita in Tehran* (New York: Random House, 2004), 109.

Index